Using SPSS
For Windows

Susan B. Gerber
Kristin Voelkl Finn

Using SPSS
For Windows

Data Analysis and Graphics

Second Edition

With 105 Figures

 Springer

Susan B. Gerber
State University of New York
Graduate School of Education
Buffalo, NY 14260
USA
gerber@buffalo.edu

Kristin Voelkl Finn
Canisius College
Graduate Education
and Leadership Department
Buffalo, NY 14208
USA
finnk@canisius.edu

SPSS is a registered trademark of SPSS, Inc.
Library of Congress Cataloging-in-Publication Data
Gerber, Susan B.
    Using SPSS for Windows: data analysis and graphics / Susan B. Gerber, Kristin Voelkl
Finn.—2nd ed.
        p. cm.
    Finn's (Voelkl) name appears first on the earlier edition (c1999).
    Includes bibliographical references and index.
    ISBN 0-387-40083-4 (alk. paper)
        1. SPSS for Windows. 2. Social sciences—Statistical methods—Computer programs. I. Finn,
Kristin Voelkl. II. Finn, Kristin Voelkl. Using SPSS for Windows. III. Title.

HA32.V63 2005
519.5′0285′53—dc22   2004065970

ISBN-10: 0-387-40083-4   Printed on acid-free paper.
ISBN-13: 978-0387-40083-9

Printed in the United States of America. (CC/HAM)

9 8 7 6 5 4 3 2 1   SPIN 10929667

springeronline.com

Preface

This book is a self-teaching guide to the SPSS for Windows computer application. It is designed to be used with SPSS version 13.0, although many of the procedures are also applicable to earlier versions of SPSS. The step-by-step format of this manual "walks" the reader through numerous examples, illustrating how to use the application. The results produced in SPSS are shown and discussed in most examples. Each chapter demonstrates statistical procedures and provides exercises to reinforce the text examples.

This book may be used in two ways – as a stand-alone manual for a student learning to use SPSS for Windows or in a course together with a basic statistics text. As a stand-alone manual, it is assumed that the reader is familiar with the basic ideas of quantitative data and statistical analysis. Thus, statistical terminology is used without providing extensive definitions. Most of the applications in this book are self-explanatory, although the reader will need to refer to a text for extensive discussion of statistical theory and procedures.

This book can also be an invaluable part of an undergraduate or graduate statistics course with a computer component and can be used easily with any elementary statistics book (e.g., *The New Statistical Analysis of Data* by Anderson and Finn, *Elements of Statistical Inference* by Huntsberger and Billingsley, *Understanding Statistics* by Mendenhall and Ott, or *Introduction to the Practice of Statistics* by Moore and McCabe). This manual provides hands-on experience with data sets, illustrates the results of each type of analysis described, and offers exercises for students to complete as homework assignments. The data sets used as examples are of general interest and come from many fields, for example, education, psychology, sociology, health, and sports. An instructor may choose to use the exercises as additional class assignments or in computer laboratory sessions. Complete answers to the exercises are available to instructors from the publisher.

Chapter 1 of this guide describes how to start the SPSS application and how to create, upload, and manipulate data files. Chapters 2 through 6 address descriptive statistics, and chapters 10 through 15 address inferential statistics. Chapters 7 through 9 discuss probability and are included primarily to illustrate the bridge between descriptive and inferential statistics. If this manual is used strictly to teach (or learn) SPSS, these chapters may not be relevant.

This manual uses SPSS for Windows, Version 13.0. System requirements include: Microsoft Windows 98, Me, NT® 4.0, 2000 or XP operating system, Pentium®-class processor, 200MB hard drive space (for the SPSS Base only), at least 128MB RAM, and an SVGA monitor. Information on installing SPSS is provided with the software. The application includes a comprehensive Help facility; the user need only click on **Help** on the main menu bar within the open application.

Information on obtaining data files used in this manual are posted on the Springer-Verlag website, at http://www.springeronline.com.

Buffalo, New York   Susan B. Gerber
Kristin Voelkl Finn

Contents

Chapter 6. Summarizing Multivariate Data: Association Between Categorical Variables 77

Part III. Probability 89

Chapter 7. Basic Ideas of Probability 91

Chapter 8. Probability Distributions 95

Chapter 9. Sampling Distributions 99

Part IV. Inferential Statistics

Part I

Introduction

Chapter 1

The Nature of SPSS

1.1 GETTING STARTED WITH SPSS FOR WINDOWS

Windows

SPSS for Windows is a versatile computer package that will perform a wide variety of statistical procedures. When using SPSS, you will encounter several types of windows. The window with which you are working at any given time is called the *active* window. Four types of windows are:

Data Editor Window. This window shows the contents of the current data file. A blank data editor window automatically opens when you start SPSS for Windows; only one data window can be open at a time. From this window, you may create new data files or modify existing ones.

Output Viewer Window. This window displays the results of any statistical procedures you run, such as descriptive statistics or frequency distributions. All tables and charts are also displayed in this window. The viewer window automatically opens when you create output.

Chart Editor Window. In this window, you can modify charts and plots. For instance, you can rotate axes, change the colors of charts, select different fonts, and rotate three-dimensional scatter plots.

Syntax Editor Window. You will use this window if you wish to use SPSS syntax to run commands instead of clicking on the pull-down menus. An advantage to this method is that it allows you to perform special features of SPSS that are not available through dialog boxes. Syntax is also an excellent way to keep a record of your analyses.

To start an SPSS session, select SPSS from the programs submenu on the Windows Start menu. Figure 1.1 shows what the screen will look like when SPSS for Windows first opens.

The Main Menu

SPSS for Windows is a menu-driven program. Most functions are performed by selecting an option from one of the menus. We refer to these menus as "pull down" menus since an entire menu of options appears when one is selected. The main menu bar is where most functions begin, and is located at the top of the window (see Fig. 1.1). Any menu may be activated by simply clicking on the desired menu item, or using the Alt-letter keystroke (each menu uses the first letter in the menu word). For example, to activate the file menu, either click the mouse on **File** or use the keyboard with **Alt-F**. The main menu bar lists 10 menus:

File. This menu is used to create new files, open existing files, read files that have been created by other software (e.g., spreadsheets or databases), and print files.

Edit. This menu is used to modify or copy text from output or syntax windows.

View. This menu allows you to change the appearance of your screen. You can, for instance, change fonts, customize toolbars, and display data using their value labels.

Data. Use this menu to make temporary changes in SPSS data files, such as merging files, transposing variables and cases, and selecting subsets of cases for analyses. Changes are not permanent unless you explicitly save the changes.

Transform. The transform menu makes changes to selected variables in the data file and computes new variables based on values of existing variables. Transformations are not permanent unless you explicitly save the changes.

Analyze. Use this menu to select a statistical procedure to be performed such as descriptive statistics, correlations, analysis of variance, and cross-tabulations.

Graphs. This menu is used to create bar charts, pie charts, histograms, and scatter plots. Some procedures under the Analyze menu also generate graphs.

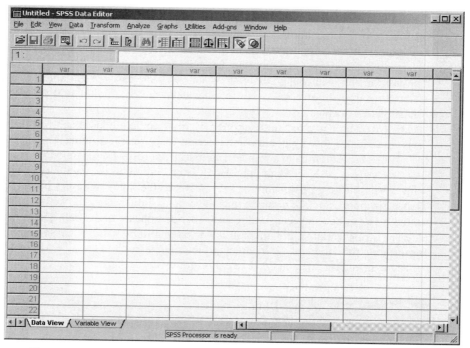

Figure 1.1 SPSS Data Editor

Utilities. This menu is used to change fonts, display information on the contents of SPSS data files, or open an index of SPSS commands.

Window. Use the window menu to arrange, select, and control the attributes of the SPSS windows.

Help. This menu opens a Microsoft Help window containing information on how to use many SPSS features.

1.2 MANAGING DATA AND FILES

Entering and selecting data files in SPSS for Windows is quite easy. We will demonstrate how to enter raw data from scratch and how to open existing data files.

Entering Your Own Data

Raw data may be entered in SPSS by using the SPSS data editor. (ASCII data may also be entered with another editor, which are then read by SPSS using the

Text Import Wizard as described below.) The SPSS editor looks like a spreadsheet or grid and is automatically opened each time you start an SPSS session. The editor is used to enter, edit, and view the contents of your data file. If you are opening an existing data file, the data will appear in the editor and you may then use the editor to change the data or add or delete cases or variables. If you are starting from scratch and wish to enter data, the data editor will be empty when it is first opened.

The data editor is a rectangle defined by rows and columns. Each cell represents a particular row-by-column intersection (e.g., row 1, column 3). All data files in the data editor have a standard format. Each row of the data editor represents a case (e.g., subject #1 or John Doe). Each column in the data editor represents a variable (e.g., heart rate or sex). Cells in the editor may not be empty. That is, if the variable is numeric and there is no valid value, the cell is represented by a "system-missing" value and a period appears in the cell. If the variable is a string variable, a blank cell is considered to be valid. (See Section 1.5 for further information on the treatment of missing values.)

To begin entering data in the data editor, follow these steps:

1. Click on **File** from the menu bar.

2. Click on **New** and then **Data** from the file pull-down menu.

3. Click on the cell in which you wish to enter data (or use the arrow keys to highlight the cell). Begin at the uppermost left cell in the rectangle. This is row 1, column 1. Once you have clicked on that cell, a darkened border will appear around the cell; this tells you that this is the cell you have selected.

4. Type in the value you wish to appear in that cell and then press Enter. You should notice that the value you type will appear at the top of the data editor window and in the cell. Notice that entering a value in this first column and pressing Enter automatically creates a variable with the default name VAR00001, which appears at the top of the column. Later we will demonstrate how to specify original names and alternate formats for variables. As an example, suppose that you are recording ages for 25 people. If the age of the first person is 18, enter 18 in the first cell.

5. Type in another value for the second case. This cell is directly below the previous cell. This location is row 2, column 1. Again, you will see the value at the top of the data editor and in the cell. Suppose the age of the second person was 22, enter 22 in row 2 column 1.

6. Repeat this process until you have entered all of the data you wish for column 1 (values for all cases on variable 1).

7. When you are ready to add another variable, click on the first cell in the next column (row 1, column 2). Suppose that "shoe size" is the next variable, and the first person has size 7. Enter this value and press enter. This will automatically create a new variable and call it VAR00002.

8. Repeat this process for all values in column 2.

9. Continue this procedure until you have entered values for all cases and variables that you wish for your data file.

Once you have entered data in the data editor, you may change or delete values. To change or delete a value in a cell, simply click on the cell you wish to alter. You will notice that a dark border appears around the selected cell, and the value in the cell appears at the top of the data editor. If you are changing the value, simply type the new value and press enter. You should see the new value replace the old value in the cell.

Adding Cases and Variables

To insert a new case (row) between cases that already exist in your data file:

1. Point the mouse arrow and click on the row number *below* the row where you wish to enter the new case. The row should be highlighted in black.
2. Click on **Data** on the menu bar.
3. Click on **Insert Cases** from the pull-down menu.

A new row is now inserted and you may begin entering data in the cells. Notice that before you enter your values, all of the cells have system-missing values (represented by a period).

To insert a new variable (column) between existing variables:

1. Click on the column variable name that is to the *right* of the position where you wish to enter a new variable. The column should be highlighted in black.
2. Click on **Data** on the menu bar.
3. Click on **Insert Variable** from the pull-down menu.

A new variable (column) is now inserted and you may begin entering data in the cells.

Deleting Cases and Variables

To delete a case:

1. Click on the case number that you wish to delete.
2. Click on **Edit** from the menu bar.
3. Click on **Clear**.

The selected case will be deleted and the rows below will shift upward.

To delete a variable:

1. Click on the variable name that you wish to delete.
2. Click on **Edit** from the menu bar.
3. Click on **Clear**.

The selected variable will be deleted and all variables to the right of the deleted variable will shift to the left. Deleting variables can also be accomplished using SPSS syntax (see Section 1.6) with the Drop and Keep subcommands.

Defining Variables

By default, SPSS assigns variable names and formats to all variables in the SPSS data file. By default, variables are named VAR##### (prefix VAR followed by five digits) and all values are valid (blanks are assigned system-missing values). Most of the time, however, you will want to customize your data file. For example, you may want to give your variables more meaningful names, provide labels for specific values, change the variable formats, and assign specific values to be regarded as "missing." To do any or all of these:

1. First, make sure that your data file window is the active window and click on the variable name that you wish to change.
2. Click on the **Variable View tab** or else double-click on the variable name in the data editor.
3. Type the name of the variable in the Name column. Variable names have to be unique, begin with a letter, and cannot contain blank spaces.
4. If you wish to change the type or format of a variable, click the button in the Type cell to open the Variable Type dialog box. By default, all variables are numeric, but you may work with other types such as names, dates, and other non-numeric data. Suppose you have a variable that contains letters (e.g., student names). This is known as a string variable and you would indicate this by clicking on **String** in the Variable Type dialog box and then clicking on **OK**.
5. Suppose you have a variable representing average cost of groceries per person that was entered to the nearest cent (e.g., 32.24) and you want to change this format so that the average cost is displayed as a whole number (rounded to the nearest dollar, e.g., 32). To change the format of the numeric variable, click in the **Width** box. The number in this box tells you the total number of columns that the variable occupies in the data file (including one column for decimal places, plus, or minus signs). For example, 8 indicates that the variable is 8 columns wide. Use the arrows to adjust the variable's column width. If you wish to change the number of decimal places, click in the **Decimals** box. The number in this box tells you how many numbers appear after the decimal place. For example, the number

32.24 would have a "width" of 5 and a 2 in the "decimal places" box. The number 32 would have a width of 2 and a 0 in the decimal places box. Use the arrows to adjust the number of decimal places.

6. If one of your variables is categorical, you can assign numbers to represent the categories of the variable. For example, the variable sex will have 2 categories: male and female. Males may have the assigned value "1" and "2" represents females. It is useful to have descriptive labels assigned to the values of 1 and 2 so that it is easy to see which number represents which category in your output files.

To assign value labels to the variable, click the button in the Values cell to open the Value Labels dialog box. Type the number representing the first category (e.g., 1) in the Value box. Type the corresponding value label (e.g., male) in the Value Label box. Click on the **Add** button. Go back to the Value box and type in the next value (e.g., 2). Type the value label for this value in the Value Label box (e.g., female), and click on **Add**. Note that each time you click Add, you will see the value and its' corresponding label appear in the window to the right of the Add button. When you have added all of the values and labels, click on **OK**.

7. If there are specific values that you would like to be treated as missing values, click on **Missing** to open the Missing Values dialog box. Click on **Discrete Missing Values** to tell SPSS that you have specific values that are considered to be missing. Type the value(s) in the boxes (you may have up to three values). If you have more than three missing values, click on **Range plus one optional discrete missing value** and enter the lower and upper bounds of the discrete variable. Click **OK** when you have entered in all of your missing values.

Opening Data Files

SPSS for Windows can read different types of data already entered into computer files. The file type we will use in this manual is the SPSS data file. These files are easily identified because (by default) each file name is followed by an ".sav" extension. SPSS data files are unique because they contain the actual data as well as information about the data such as variable names, formats, and column locations. These files are written in a special code that is read and interpreted by the SPSS program. If you try to read these data files with software other than SPSS, the file will look like lines of secret code and will not make sense to you. However, they make a great deal of sense to SPSS, and this is why reading them with SPSS is so easy. If you would like to look at the information contained in an SPSS data file (that is currently open), you can do this by clicking on **File** in the menu bar, and then choose **Display Data File Information** and then **Working File**.

SPSS for Windows can also read raw data that are in simple text files in standard ASCII format. Text files are usually identified by a ".dat" or ".txt" extension. These are data files that just contain ordinary numbers (or letters). There is no additional information contained in the file such as variable locations, formats, labels, missing values, etc. (SPSS .sav data files do contain this additional information). You can read text files with many different software programs, including WordPad. SPSS can read text data files that are formatted as fixed or tab-delimited.

The SPSS Data Editor is designed to read a variety of formats in addition to SPSS data files and ASCII text files. For example, spreadsheet files created with Lotus 1-2-3 and Excel, database files created with dBASE and SQL formats, and SYSTAT data files.

Reading SPSS Data Files

We will illustrate how to read an existing SPSS data file. The reader may follow along using the data accompanying this guide.

To open a data file:

1. Click on **File** from the menu bar.
2. Click on **Open** on the file pull-down menu.
3. Click on **Data** on the open pull-down menu. This opens the Open File dialog box as shown in Figure 1.2.
4. Choose the correct directory from the **Look in:** box at the top of the screen.
5. Point the arrow to the data file you wish to open and click on it. By default, all SPSS data files (*.sav) in the current directory will be displayed in the list. If your data file is not visible in the file name box, use the left and right arrows to scroll through the files until you locate your desired file. Note that all of the SPSS data files have the .sav extension, and this is designated in the Files of type window. Before you open a data file, make certain that the file type is correct. If you are reading SPSS data files and the file type box does not read "SPSS (*.sav)," you must scroll through the file types and select that type. For example, to open the file called "football.sav," highlight the name of this file by clicking on it with the mouse button.
6. Click on **Open**. You should now see the contents of the data file displayed in the Data Editor window. The "football.sav" data file contains two variables, "height" and "weight," for 56 football players from Stanford University. The variable names are displayed at the top of the Data Editor; each column contains one variable. The rows in the data file are the cases; in this data file there are 56 cases.

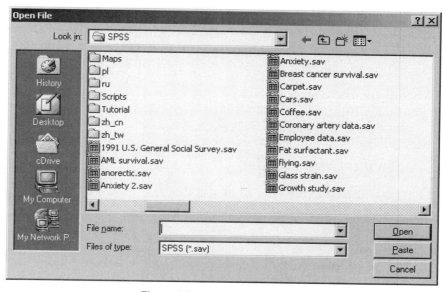

Figure 1.2 Open File Dialog Box

Note: Most of the examples in the following chapters use the SPSS data files that are provided with this manual. Unless you are required to enter data on your own into a new file, all procedures assume that you have opened the SPSS data file before beginning any computations or analyses.

Reading Data Files in Text and Other Formats

To read a text data file, begin at the main menu bar in the Data Editor window:

1. Click on **File.**
2. Click on **Read Text Data**.
3. Select the appropriate file from the Open file dialog box and click **Open**.
4. Follow the steps in the **Text Import Wizard** to read the data file. You will have to answer questions about type of data, arrangement of data, number of cases to import, and missing values. Use the Help button of the Text Import Wizard for more detailed information.

To open data from a file such as an Excel spreadsheet, begin at the Data Editor window:

1. Click on **File.**
2. Click on **Open** and then click on **Data.**

3. Select the file format from the drop-down list of file types in the **Files of type:** box.

4. Choose the appropriate directory and file.

5. Click on **Open**.

Excel, Lotus, and SYLK variable names are read from the file and appear in the first row of the spreadsheet. If the spreadsheet does not contain variable names, SPSS provides default names using column letters.

Saving Data Files

Unless you save your files, all of your data and changes will be lost when you leave the SPSS session. To save a file, first make the Data Editor the active window. Then:

1. Click on **File** from the menu.

2. Select **Save** from the list of options in the File pull-down menu.

3. Select the appropriate directory in the **Save in:** box. Type the name of your file in the **File name** box. Notice that the default file type is set for SPSS format as indicated by the ".sav" extension.

4. Click on **Save**.

By default, this will save the data file as an SPSS data file. If you were working with a previously existing data file, the old file will be overwritten by the modified data file. To save the file with a different name, select **Save As ...** from the File pull-down menu. Note that it is always recommended that you preserve your original data in a separate file in case you ever need to return to it.

If you wish to save the data file in a format other than SPSS (e.g., Lotus, Excel, dBASE, fixed-format ASCII text):

1. Click on **File** from the menu.

2. Select **Save As** from the list of options in the File pull-down menu.

3. Select the appropriate directory in the Save in: box. Type the name of your file in the **File name** box.

4. Choose the appropriate file type in the **Save as type:** box.

5. Click on **Save**.

1.3   *TRANSFORMING VARIABLES AND DATA FILES*

At times, you may need to alter or transform the data in your data file to allow you to perform the calculations you require. There are many ways in which you can transform data. This section discusses three commonly used techniques: computing new variables, recoding variables, and selecting subsets of cases.

Computing New Variables

There may be occasions when you need to compute new variables that combine or alter existing variables in your data file. For instance, your data file may contain daytime and nighttime sleeping hours for a sample of infants, but you are interested in examining total sleep hours (i.e., the sum of the separate daytime and nighttime hours).

To create a new variable:

1. Click on **Transform** from the menu bar.
2. Click on **Compute** from the pull-down menu. This opens the Compute Variable dialog box (see Fig. 1.3).
3. Enter the name of the new variable (in the above illustration, total) in the Target Variable box. (You also have the option to describe the nature and format of the new variable by clicking on the **Type & Label** box.)
4. You will then need to perform a series of steps to construct an expression used to compute your new variable. In this illustration, you would first select the daytime variable ("daysleep") from the variable list box on the left-hand side of the dialog box and move it to the Numeric Expression box using the **right directional arrow.**
5. Then click on the "+" from the calculator pad. You will notice that a plus sign is placed in the Numeric Expression box after the word daytime.
6. Complete the expression by selecting the nighttime variable ("nightsleep") and moving it to the Numeric Expression box, following the instructions in step (4) above.
7. When you have completed the expression, click on **OK** to close the Compute Variable dialog box. Your new variable will be added to the end of your data file.

In addition to simple algebraic functions on the calculator pad (+, -, x, ÷), there are many other arithmetic functions such as absolute value, truncate, round, square root, and statistical functions including sum, mean, minimum, and maximum. These are displayed in the Function group box to the right of the calculator pad. First, select a procedure in the Function group window, and then select the specific function in the Functions and Specific Variables window.

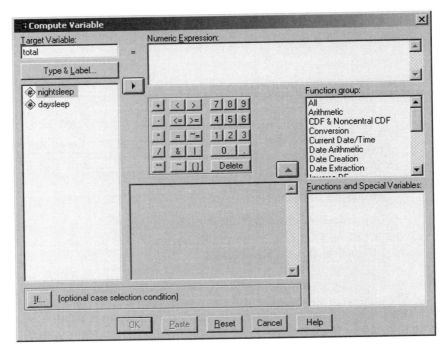

Figure 1.3 Compute Variable Dialog Box

Recoding Variables

Recoding variables is often a useful technique. There are several instances in which you may need to modify the values in your original data file. You can recode either categorical or numeric variables. For instance, you may have a data file containing information about Olympic athletes such as number of medals won, years of training, and type of sport. A "sport" variable, with a value of 1 to 100, indicates the type of event (e.g., bobsled, diving, etc). Suppose that you are interested in examining the medal distribution based on whether athletes competed in the summer or winter Olympics. For this, you would need to recode the sport variable from 100 categories to two.

Or, you may want to categorize a discrete or continuous numeric variable into a limited number of groups. For example, your data file may contain the number of years of training and you wish to group them into three categories: 1–5 years, 6–10 years, and more than 10 years.

You have two options available for recoding variables. You may recode values into the same variable, which eliminates all record of the original values. This is a useful function for correcting obvious data errors, changing system missing values into a valid value for profiling item nonrespondents, or collapsing a number of values when only a few cases responded in a particular way and a meaningful assessment of these few cases cannot be conducted. You also have the option to create a new variable containing the recoded values. This preserves the values of the original variable. If you think that there may be a reason that you would need to have record of the original values, you should select the second option.

Recoding into the Same Variable

To recode into the same variable:

1. Click on **Transform** from the main menu.
2. Click on **Recode** from the pull-down menu.
3. Click on **Into Same Variables** to open the Recode into Same Variable dialog box.
4. Select the name of the variable to be recoded, and move it to the Variables box with the **right arrow button**.
5. Click on **Old and New Values**. This opens the Old and New Variables dialog box (see Fig. 1.4).
6. For each value (or range of values) you want to recode, you must indicate the old value and the new value, and click on **Add**. The recode expression will appear in the **Old --> New** box.

Old values are those before recoding — the values that exist in the original variables. There are several alternatives for recoding old values, including the Value and the Range options discussed below.

Value Option. You may use the Value option for cases in which you want to collapse existing categorical variables such as the "sport" illustration above. In your original data file, you have an existing value (1 through 100) for each type of sport, but you need to recode these values into the values 1 and 2 — 1 representing summer and 2 representing winter. For example:

Original		Recoded	
Category	Value	Category	Value
Diving	1	Summer	1
Gymnastics	2	Summer	1
Skating	3	Winter	2
Bobsled	4	Winter	2

The steps to perform this recoding are straightforward:

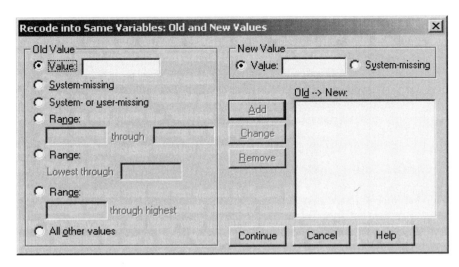

Figure 1.4 Recode into Same Variables: Old and New Values Dialog Box

1. Type **1** in the **Value** box of the **Old Value** section, indicating the existing value for Sport 1 (Diving).

2. Type a **1** in the **Value** box of the **New Value** section, indicating that it is to be recoded as season 1 (Summer).

3. Click on **Add**. You will notice that the expression 1 --> 1 will appear in the **Old -->New** box.

Follow the same procedure for the rest of the values of the original variable. For example, you would recode the old value of 2 to the new value of 1, and click on **Add**. Note that because Diving was coded 1 both before and after recoding, it was not necessary to include it in the recode procedure. Doing so, however, may assist you in making sure that you have included all values to be recoded.

Range Option. You may also recode variables using the Range option. This is most useful for numerical variables. The procedure is similar to that discussed above. To recode the years of training in this example, the existing values would be recoded as follows:

Original Range of Values	Recoded Value
Lowest through 5	1
6 through 10	2
11 through highest	3

Instead of choosing the value option in the old value section, you may use the range option as follows:

1. Type **5** in the **Range: Lowest through** ___ box ; the middle range option box.

2. Type **1** in the **Value** box under the **New Value** section.

3. Click on **Add**. The expression will appear in the **Old --> New** box.

4. Type **6** and **10** in the two boxes of the first range option: **Range:** ___ **through** ___.

5. Type **2** in the **Value** box under the **New Value** section.

6. Click on **Add**.

7. Type **11** in the **Range:** ___ **through highest** box; the bottom range option of the Old Value section.

8. Type **3** in the Value box under the **New Value** section.

9. Click on **Add**.

When you have indicated all the recode instructions, using either the Value or Range method, click on **Continue** to close the Recode Into Same Variables: Old and New Values dialog box. Click on **OK** to close the Recode Into Same Variables dialog box. While SPSS performs the transformation, the message "Running Execute" appears at the bottom of the application window. The "SPSS Processor is Ready" message appears when transformations are complete.

Recoding into Different Variables

The procedure for recoding into a different variable is very similar to that for recoding into the same variable:

1. Click on **Transform** from the main menu.

2. Click on **Recode** from the pull-down menu.

3. Click on **Into Different Variables** to open the Recode into Different Variable dialog box.

4. Select the name of the variable to be recoded from the variables list, and move it to the Input Variable --> Output Variable box with the **right arrow button**.

5. Type in the name of the new variable you wish to create in the Output Variable box. If you wish, you may also type a label for the variable.

6. Click on **Change**, and the new variable name will appear linked to the original variable in the Input Variable --> Output Variable box.

7. Click on **Old and New Values**. This opens the Old and New Variables dialog box.

8. The procedure for identifying old and new values is the same as that discussed in the Recoding into the Same Variable subsection, with one exception. Because you are creating a new variable, you must indicate new values for *all* of the old values, even if the value does not change. (This is optional when recoding to the same variable.) Because this step is mandatory, SPSS provides a **Copy Old Value(s)** option in the New Value box.

9. When you have indicated all the recode instructions, click on **Continue** to close the Recode Into Different Variables: Old and New Values dialog box.

10. Click on **OK** to close the Recode Into Different Variables dialog box. While SPSS performs the transformation, the message "Running Execute" appears at the bottom of the Application Window. The "SPSS Processor is Ready" message appears when transformations are complete, and a new variable appears in the data editor window. The new variable is added to your data file in the last column displayed in the Data Editor window.

Selecting Cases

There may be occasions when you need to select a subset of cases from your data file for a particular analysis. You may, for instance, have a data file containing height and weight for 200 individuals, but you need to know the average height of individuals over 120 pounds. Or, you may simply wish to select a random sample of cases from a very large data file.

To select subset of cases:

1. Click on **Data** from the main menu.

2. Click on **Select Cases** from the pull-down menu. This opens the Select Cases dialog box (see Fig. 1.5).

There are several ways in which you can select a subset from a larger data file. We will discuss the If Condition and Random Sample methods.

If Condition

In the height and weight example given above, you would need to:

1. Select the **If condition is satisfied** option and click on **If** to open the Select Cases: If dialog box.

2. Select the weight variable from the variable list box, and move it into the box above the calculator pad with the **right arrow button**.

3. Using the calculator pad, click on >. This sign will appear in the upper box.

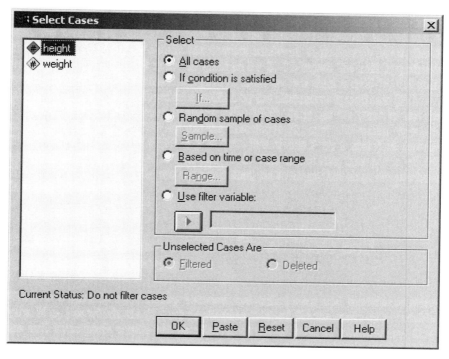

Figure 1.5 Select Cases Dialog Box

4. Using the number pad, Click on **1**, then **2**, then **0**, to create the expression "weight > 120" in the upper right-hand box.

5. Click on **Continue** to close the Select Cases: If dialog box.

Random Sample

Another method for selecting subcases is to choose a random sample from your data file:

1. From the Select Cases dialog box (see Fig. 1.5, above) select the **Random sample of cases** option and click on **Sample** to open the Select Cases: Random Sample dialog box.

2. Type either a percentage or a precise number of cases in the appropriate box.

3. Click on **Continue** to close the Select Cases: Random Sample dialog box.

You should now be back at the Select Cases dialog box. Click on **OK** to close this dialog box. The message "Running Execute" will appear as SPSS processes your instructions. SPSS creates a new "filter" variable with the value 1 for all cases selected and 0 for all cases not selected. There is also a "Filter On" message at the bottom of the application window to remind you that your subsequent analyses will be performed only on the designated subset of data. Furthermore, there are diagonal lines through the row line numbers of the Data Editor window for all cases not selected. SPSS uses the filter variable to determine which cases to include in subsequent analyses. Unselected cases remain in the data file but are not included in subsequent analyses. You can use the unselected cases by turning the filter off. To turn the filter off, go back into the Select Cases dialog box and click on **All Cases** in the Select box, and then click on **OK**.

1.4   MISSING VALUES

In many situations, data files do not have complete data on all variables, that is, there are missing values. You need to inform SPSS when you have missing values so that all computations are performed correctly. With SPSS, there are two forms of missing values: system-missing and user-defined missing.

System-missing values are those that SPSS automatically treats as missing (without the user having to explicitly inform SPSS). The most common form of this type of value is when there is a "blank" in the data file. For example, the data value for a person is missing if the information was not provided. A period is displayed in the data file cell that does not have a value. When SPSS reads this variable, it will read a blank, and thus treat the value as though it is missing. Any further computations involving this variable will proceed without the missing information. For instance, suppose you wish to calculate the average amount of daily water consumption for a sample of 20 adults, but you only have data entered for 19 people. SPSS will read the "valid" values for the 19 adults, ignore the missing value, and compute the average based on the 19 individuals.

User-defined missing values are those that the user specifically informs SPSS to treat as missing. Rather than leaving a blank in the data file, numbers are often entered that are meant to represent missing data. For example, suppose you have an age variable that ranges from 1 through 85. You could use the number 99 to represent those individuals who were missing information on age. (You could not use any numbers from 1 to 85 since these are all valid values.)

In this example, you need to inform SPSS that the value 99 is to be treated as a missing value, otherwise it will treat is as valid. This is explained in Section 1.2, but in brief, you need to do the following: switch to Variable view in the data editor and click the "..." button in the **Missing** column. Enter 99 in one of the **Discrete Missing Values** boxes and click on **OK**. When SPSS reads this variable, it will then treat 99 as a missing value and not include it in any computations involving the "age" variable. User-missing values will look like valid values in the data editor, but are internally flagged as "missing" in SPSS data files, and labeled as missing in the output for some procedures.

Most SPSS computations will display the valid number of cases in the output. This is the number of cases that were not system-missing and/or user-defined missing; these cases were used in the computations. The number of missing cases (not used in the computations) is typically displayed as well.

Analyses with Missing Data

When you have missing data, they can be treated in several ways. Missing data is a complex issue and can be problematic. If you do not specify how to handle missing data in some analyses, cases that have missing values for any of the variables named in the analysis are omitted from the analysis using pairwise or listwise deletion. For example, suppose you wish to calculate the correlations between three sets of achievement scores: math, science, and reading. Three correlations will be computed: math with science, math with reading, and science with reading. Using **pairwise deletion**, cases that do not have valid scores for both measures (e.g., math and science) will be excluded from the computation. Using **listwise deletion**, cases that do not have valid scores on all measures (i.e., math, science, and reading) will be excluded from the computations. For example, suppose you have a data file containing 1,000 cases, and all cases have a valid math score, 700 cases have a valid science score, and 400 cases have a valid reading score. Using pairwise deletion, the correlation between math and science will be based on the 700 cases have complete data on both achievement measures. Using listwise deletion, the same correlation will be based on only 400 cases because only 400 cases have complete data on all three measures. Further, any correlation you calculate with this sample would be based on 400 cases if you specify listwise deletion. As you can see, listwise deletion can greatly reduce the sample size used in your analyses. On the other hand, listwise deletion ensures that all computations are based on the same number of cases.

By default, SPSS uses either pairwise or listwise deletion depending on the procedure. For example, listwise deletion is the default for multiple regression, but pairwise deletion is the default for bivariate correlations. To designate pairwise or listwise deletion, click on the **Options** button after opening the dialog box for the appropriate analysis and then choose the method for handling missing values and click **Continue**.

In some situations, you may wish to substitute missing values with a new value to be used in the analysis rather than excluding cases from your computations. For example, in repeated measures analyses, complete data in the series is required for the analysis, so missing data are not allowed. Substituting missing values can be done with SPSS using the **Replace Missing Values** function, which creates new variables from existing ones, replacing missing values with estimates created from one of several procedures. One of the most common procedures is to replace missing values with the mean for the entire series. Other procedures include replacing missing values with the mean or median of nearby points (as indicated by the user) or linear interpolation. Overall, missing data represents a complex problem for the data analyst and simple solutions such as replacement of missing values is generally not advised. Further, imputing missing values is not recommended for variables with a large amount of missing data.

To replace missing values with the mean for the entire series:

1.   Click on **Transform** from the main menu.

2.   Click on **Replace Missing Values** to open the Replace Missing Values dialog box (see Fig. 1.6).

3.   Click on the variable for which you wish to substitute values and click on the **right arrow button** to move it into the New Variable(s) box. By default, a new variable name will be created by using the first six characters of the existing variable followed by and underscore and a sequential number. For example, the variable "math" would be replaced with a new variable "math_1." The new variable name will appear in the Name: field in the Name and Method box.

4.   Click on **Series Mean** in the Method: box.

5.   Click on **OK**.

The output shows the number of cases for which the mean math score for the series was substituted for missing math scores.

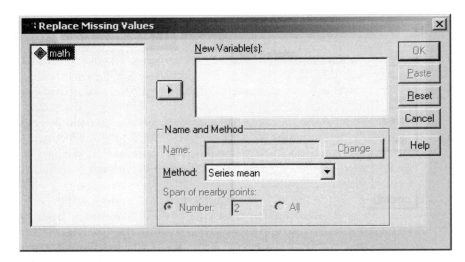

Figure 1.6 Replace Missing Values Dialog Box

1.5   EXAMINING AND PRINTING OUTPUT

After running a procedure, SPSS results are displayed in the output Viewer window (see Fig. 1.7). From this window, you can examine your results and manipulate output. The viewer is divided into two panes. An outline of the output contents is arranged on the left side. The right side contains the detailed output of your procedures such as descriptive statistics, frequency distributions, results of *t*-tests, as well as charts and plots including histograms, box-and-whisker plots, and scatter plots. Each time you direct SPSS to create a chart or graph, it displays it in the viewer. Double-click on a graph or chart if you wish to edit it. You can go directly to any part of the output by clicking on the appropriate icon in the outline in the left pane. You may also view the output by using the directional arrows buttons on the vertical scroll bar at the right edge of the window.

Printing Output

To print the contents of an Output Window:

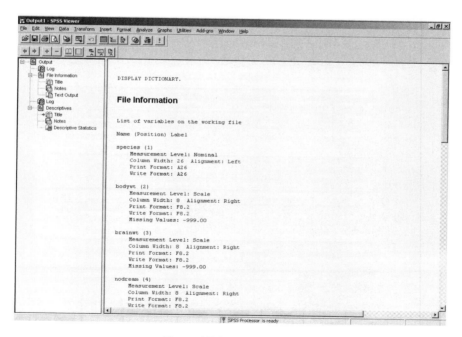

Figure 1.7 Viewer window

1. Make the viewer window the active window.
2. Click on **File** from the main menu.
3. Click on **Print** from the pull-down menu. This opens the Print dialog box.
4. If you wish to print the entire file, click on the **All Visible Output** option. If you wish to print only a selected block, click on the **Selection** option. To print only a section of the file, you need to use the "click-and-drag" method to highlight the area before opening the Print dialog box.
5. Click on **OK** to close the Print dialog box and print the file.

1.6   USING SPSS SYNTAX

As illustrated throughout this book, most SPSS procedures are conducted using the pull-down menus because they are convenient and easy to use. However, an alternative way to run SPSS procedures is through command syntax. SPSS commands are the instructions that you give the program for conducting procedures. SPSS syntax commands are typed into a command file using the SPSS syntax editor. Syntax files have the extension ".sps."

There are several reasons why command syntax is useful, such as when the user wants to: (1) have a record of the analyses conducted during a session; (2) repeat long and complex analyses; (3) review how variables were created or transformed; and (4) modify commands to run slightly different or customized statistics.

When working with syntax, the user must enter commands instructing the program what procedures to conduct. You can enter syntax by either typing or pasting syntax into the syntax editor. Because most users do not know the commands from memory, it is useful to refer to the *SPSS Syntax Reference Guide* for a complete reference to the command syntax. Help is also available by using the Help button on the toolbar in the syntax editor window. Pasting syntax commands from dialog boxes is perhaps the easiest way to construct syntax commands. Rather than typing the commands, you initiate a procedure using pull-down menus and then instruct SPSS to provide the commands and paste them into the syntax editor.

To open a new window and begin typing commands:

1. Click on **File** from the main menu.
2. Click on **New** from the pull-down menu.
3. Click on **Syntax** to open the SPSS syntax editor (see Fig. 1.8).
4. Begin typing syntax into the editor.

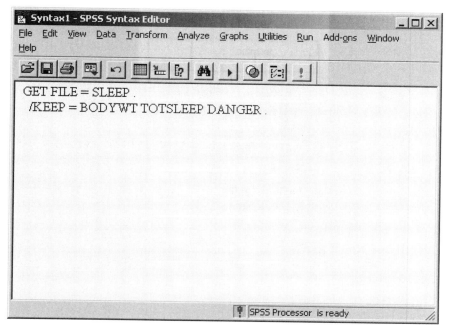

Figure 1.8 SPSS Syntax Editor

For example, suppose you want to open the sleep.sav data file, but you only want to read a subset of variables — body weight, total sleep, and danger index. The syntax command would be:

GET FILE = SLEEP .
 /KEEP = BODYWT TOTSLEEP DANGER .

You can also run a procedure by pasting syntax from a dialog box. When you use the paste button, SPSS creates the syntax commands to execute procedures requested from pull-down menus. For example, to compute a new variable (total sleep hours) as shown in Section 1.3, follow steps 1–6. Instead of clicking on **OK**, click on the **Paste** button. The compute commands will automatically be displayed in a syntax window. To run the syntax commands, click the **Right arrow button** on the toolbar.

Once you have created a syntax file, you can save it using the same procedures described in Section 1.2 of this chapter. The file can then be opened and edited for future modifications. Make sure when you open, edit, and save a syntax file that you correctly identify it with the ".sps" file type.

Chapter Exercises

1.1 Select the data file "football.sav" and without opening the file, answer the following:

    a. How many variables are in the file?

    b. What is the format for the weight variable?

1.2 Open the SPSS data file "spit.sav" and answer the following:

    a. Is this a text file?

    b. How many cases are in the file?

    c. How many variables are in the file?

    d. Are there any missing data?

1.3 Enter the age and sex for 10 students in your class into a new data file using the SPSS data editor.

    a. Save the data as an SPSS data file.

    b. Save the data as an ASCII data file.

    c. Once you have saved and exited the file, re-open the ASCII data file and enter a new variable named "minors" with the value of 1 for students under the age of 19 and 2 for students 19 or older. Save the file as an SPSS file.

    d. Re-open the SPSS data file and delete the fifth case. Was this a minor?

Part II

Descriptive Statistics

Chapter 2

Summarizing Data Graphically

A statistical data set consists of a collection of values on one or more variables. The variables can be either numerical or categorical. Numerical variables are further classified as discrete or continuous. These distinctions determine the statistical approaches that are appropriate for summarizing the data. Examples of data include

- • crime rates for large cities across the United States;
- • body temperatures for a randomly chosen sample of adults;
- • the numbers of errors made by cashiers on an 8-hour shift;
- • the gender of individuals purchasing tickets to a concert; and
- • occupations of a sample of fathers and their sons.

One approach to organizing data is through a chart or graph. The type of chart you use depends in part on the way the data are measured — in categories (e.g., occupations) or on a numerical scale (e.g., number of errors). This chapter demonstrates how to examine different types of data through frequency distributions and graphical representations. Section 2.1 describes methods for summarizing categorical data, while Section 2.2 pertains to discrete and continuous numerical variables.

2.1   SUMMARIZING CATEGORICAL DATA

Categorical variables are those that have qualitatively distinct categories as values. For example, gender is a categorical variable with categories "male" and "female." Information on the coding and labeling of categorical data is given in Chapter 1.

Frequencies

One way to display data is in a frequency distribution, which lists the values of a variable (e.g., for the variable occupation: professional, manager, salesperson, etc.) and the corresponding numbers and percentages of participants for each value.

Let us begin by creating a simple frequency distribution of occupations using the "socmob.sav" SPSS data file on the website accompanying this manual. Follow along by using SPSS to open the data file on your computer (using the procedure given in Chapter 1). This data set was used in a study of the effects of family disruption on social mobility. The study collected data on fathers' occupations, their sons' occupations, family structure (intact/not intact), and race.

Notice that the data view lists numbers as the values for all of the variables, even though the variable is a categorical variable. The use of numbers to represent categories was described in Chapter 1. To see the categories each of the values represent, you can examine the contents of the data file (variable labels, variable type, and value labels) by clicking on **Utilities** on the menu bar and clicking on **Variables** from the pull-down menu. You can also click on the **value labels** button on the toolbar, as displayed in Figure 2.1. This will display the value labels (e.g., laborer, manager, professional) in the data editor.

To create a frequency distribution of the father's occupation variable:

1. Click on **Analyze** from the menu bar.
2. Click on **Descriptive Statistics** from the pull-down menu.
3. Click on **Frequencies** from the second pull-down menu to open the Frequencies dialog box (see Fig. 2.2).

Figure 2.1 Toolbar with value labels button activated

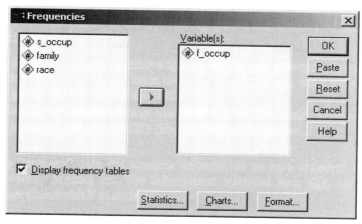

Figure 2.2 Frequencies Dialog Box

4. Click on the label/name of the variable you wish to examine ("f_occup") in the left-hand box.

5. Click on the **right arrow button** to move the variable name into the Variable(s) box.

6. Click on **OK**.

The frequency distribution produced by SPSS is shown in Figure 2.3. This figure shows the content of the output — that which is in the right-hand frame of your Output Viewer.

The "Statistics" table in the output indicates the number of valid and missing values for this variable. There are 1156 valid cases and no missing values. The "father's occupation" table displays the frequency distribution. The occupational categories appear in the left-hand column of this table. The "Frequency" column contains the exact number of cases (e.g., number of fathers) for each of the categories. For example, there are 476 fathers who are laborers and 136 fathers who are professionals.

The numbers in the "Percent" column represent the percentage of the total number of cases that are in each occupational category. These are obtained by dividing each frequency by the total number of cases and multiplying by 100. For example, 11.8% of the sample is comprised of professionals ($136/1156 \times 100$).

Statistics

father's occupation

N	Valid	1156
	Missing	0

father's occupation

		Frequency	Percent	Valid Percent	Cumulative Percent
Valid	laborer	476	41.2	41.2	41.2
	craftsperson	204	17.6	17.6	58.8
	salesperson	272	23.5	23.5	82.4
	manager	68	5.9	5.9	88.2
	professional	136	11.8	11.8	100.0
	Total	1156	100.0	100.0	

Figure 2.3 Frequency Distribution of Father's Occupation Variable

The "Valid Percent" column takes into account missing values. In this case, there are no missing values, so the "Percent" and "Valid Percent" columns are the same. The "Cumulative Percent" is a cumulative percentage of the cases for the category and all categories listed before it in the table. For example, 82.4% of all cases in the sample include laborers, craftsmen, and salesmen (41.2% + 17.6% + 23.5%, within rounding error). The cumulative percentages are not meaningful unless the scale of the variable has at least ordinal properties. Ordinal means that the values of the variable are ordered. Numerical variables have ordinal properties, as do ordinal categorical variables (e.g., a variable measuring size, with values equal to small, medium, large, and extra large). The father's occupation variable does not have ordinal properties. That is, being a salesman is not "higher" or "lower" in the list of occupations than is a laborer. The occupations could have been listed in another order without affecting the interpretation of the data.

Frequencies with Missing Data

In this data file, there are no missing cases. Suppose, however, that the families with identification numbers 10128, 10129, 10180, 10343, 10350, 10370, 10434, 10435, 10500, and 10501 were missing information on father's occupation. The frequency distribution for this altered data set would appear as in Figure 2.4. Note that there is an additional row in this distribution chart — the Missing row — which indicates that there are 10 cases for which father's occupation was not

father's occupation

		Frequency	Percent	Valid Percent	Cumulative Percent
Valid	laborer	473	40.9	41.3	41.3
	craftsperson	200	17.3	17.5	58.7
	salesperson	269	23.3	23.5	82.2
	manager	68	5.9	5.9	88.1
	professional	136	11.8	11.9	100.0
	Total	1146	99.1	100.0	
Missing	System	10	.9		
Total		1156	100.0		

Figure 2.4 Frequency Distribution for Father's Occupation with Missing Cases

coded. Note also that the Percent and Valid Percent columns now indicate different figures, because of the difference in the denominator used to compute the figures. In the case of laborers, for instance, Percent is computed as 473/1156 × 100, and Valid Percent is computed as 473/1146 × 100.

Bar Charts

A bar chart is also useful for examining categorical data. In a bar chart, the height of each bar represents the frequency of occurrence for each category of the variable. Let us create a bar chart for the occupation data using an option within the Frequencies procedure.

From the Frequencies dialog box (see steps 1–3 of the Frequencies section):

1. Click on **Charts** to open the Frequencies: Charts dialog box (see Fig. 2.5).
2. Click on **Bar charts** in the Chart Type box.
3. Choose the type of values you want to chart — frequencies or percentages — in the Chart Values box. For this example, we have selected frequencies.
4. Click on **Continue**.
5. Click on **OK** to run the chart procedure.

A bar chart like that in Figure 2.6 should appear in your SPSS Viewer.

The information displayed in this chart is a graphical version of that shown in the frequency distribution in Figure 2.3. The occupation group with the greatest number of people is laborer; the occupation group with the fewest is manager. There are 204 craftspeople; this is determined by looking at the vertical (frequency) axis.

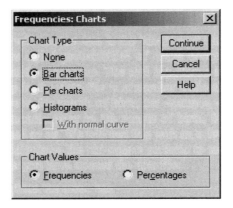

Figure 2.5 Frequencies: Charts Dialog Box

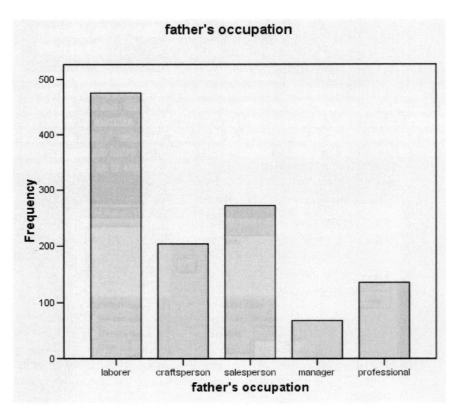

Figure 2.6 Bar Chart of Father's Occupation Variable

2.2 SUMMARIZING NUMERICAL DATA

There are two types of numerical variables — discrete and categorical. The values for discrete variables are counting numbers. For example, an American football game is won by one, two, or three points, not a quantity in between. Continuous variables, on the other hand, do not have such indivisible units. Body temperature, for instance, can be measured to the nearest degree, half-degree, quarter-degree, and so on. For practical purposes in SPSS, there is no difference in summarizing these two types of numerical data.

We shall use the data in the "football.sav"[1] data file to illustrate graphical summaries of numerical data. This file contains data on 250 National Football League (NFL) games from a recent season. One of the variables in this file is "winby," representing the number of points by which the winning team was victorious. We can create a frequency distribution of the winby variable using the same procedures as outlined in Section 2.1. The frequency distribution is in Figure 2.7. We see that 32 games were won by 3 points. This is 12.8% of the 250 games. The cumulative percent column is meaningful with numerical data, and we see that 35.2% of the games were won by 6 or fewer points (or, by "less than a touchdown").

We use histograms instead of bar charts to graphically display numerical data. There are several ways to obtain a histogram in SPSS. One such procedure is identical to the one used in Section 2.1 except you click on the histogram option in the Frequencies: Charts dialog box (Fig. 2.5). An alternative method is to use the Explore procedure, as illustrated below:

1. Click on **Analyze** on the menu bar.

2. Click on **Descriptive Statistics** from the pull-down menu.

3. Click on **Explore** from the pull-down menu. This opens the Explore dialog box as shown in Figure 2.8.

4. Click on the name of the variable ("winby") and click on the **top right arrow button** to move it to the Dependent List box. (In this example, there is no independent, or Factor, variable.)

5. Click on **Plots** in the Display box. This will suppress all statistics in the output. (If you also want SPSS to provide summary statistics, click on **Both.**)

6. Click on the **Plots button** to open the Explore: Plots dialog box.

---

[1] Appreciation for this and several other data sets used in this manual is expressed to the *Journal of Statistics Education*, http://www.amstat.org/publications/jse/, an international resource for teaching and learning of statistics.

Statistics

WINBY

N	Valid	250
	Missing	0

WINBY

		Frequency	Percent	Valid Percent	Cumulative Percent
Valid	1	9	3.6	3.6	3.6
	2	8	3.2	3.2	6.8
	3	32	12.8	12.8	19.6
	4	17	6.8	6.8	26.4
	5	9	3.6	3.6	30.0
	6	13	5.2	5.2	35.2
	7	23	9.2	9.2	44.4
	8	6	2.4	2.4	46.8
	9	5	2.0	2.0	48.8
	10	20	8.0	8.0	56.8
	11	9	3.6	3.6	60.4
	12	1	.4	.4	60.8
	13	3	1.2	1.2	62.0
	14	17	6.8	6.8	68.8
	15	10	4.0	4.0	72.8
	16	8	3.2	3.2	76.0
	17	4	1.6	1.6	77.6
	18	5	2.0	2.0	79.6
	19	6	2.4	2.4	82.0
	21	4	1.6	1.6	83.6
	22	2	.8	.8	84.4
	23	1	.4	.4	84.8
	24	6	2.4	2.4	87.2
	25	8	3.2	3.2	90.4
	26	3	1.2	1.2	91.6
	27	5	2.0	2.0	93.6
	28	5	2.0	2.0	95.6
	31	3	1.2	1.2	96.8
	32	2	.8	.8	97.6
	34	1	.4	.4	98.0
	35	1	.4	.4	98.4
	36	2	.8	.8	99.2
	38	1	.4	.4	99.6
	43	1	.4	.4	100.0
	Total	250	100.0	100.0	

Figure 2.7 Frequency Distribution of Points by Which Football Games Were Won

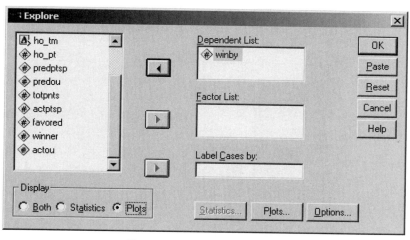

Figure 2.8 Explore Dialog Box

7. Click on **Histogram** in the Descriptive box. (In this example, we are only interested in the histogram, so we also click **None** instead of **Factor levels together** under boxplots and click off the **Stem-and-leaf** option under Descriptive.)
8. Click on **Continue**.
9. Click on **OK** to run the procedure.

The SPSS Viewer will open with the results of the procedure. They are contained in Figure 2.9.

Changing Intervals

The "winby" variable has a range of 42 points (1–43), and the histogram with 22 intervals (selected by SPSS) adequately conveys the nature of the data. It is possible, however, to edit the histogram to change the number of intervals displayed (or the interval width). For example, to change the number of intervals on the x-axis in the above histogram to 10, follow the steps below:

1. In the SPSS Viewer, double click on the histogram to open the SPSS Chart Editor.
2. Click on **Edit** from the menu bar.
3. Click on **Select X Axis** from the pull-down menu to open the Properties dialog box (Fig. 2.10).
4. Click on the **Histogram Options** tab.

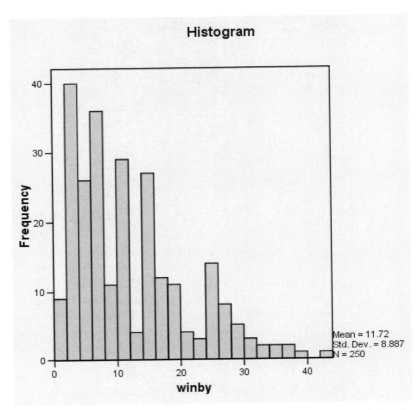

Figure 2.9 Output of Explore Procedure Creating Histogram of Number of Points by Which Football Games Were Won

5. In the Bin Sizes section, select **Custom**.

6. Change the number of intervals to 10.

7. Click on **Apply** to redraw the graph.

8. The cancel button changes to a close button. Click on **Close** to close the Properties dialog box.

9. Click the **X** in the right corner of the Chart Editor window to close it. Note that the histogram now has 10 intervals, but it is still positively skewed in shape (Fig. 2.11).

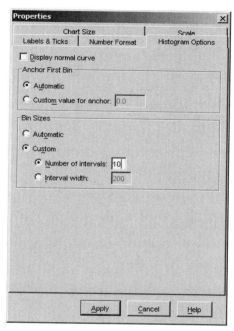

Figure 2.10 Histogram Options Tab of Properties Dialog Box

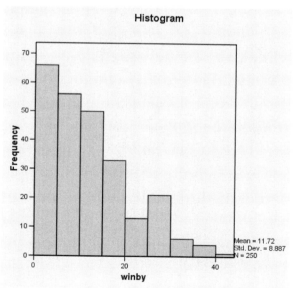

Figure 2.11 Histogram of Points by Which Games Were Won (10 Intervals)

Stem-and-Leaf Plot

Another way to display numerical data is with a stem-and-leaf plot. This display divides data into intervals. The value for each observation is split into two parts — the stem and the leaf. The stem represents the interval and the leaf represents the last digit(s) of the actual data points.

To direct SPSS to produce a stem-and-leaf plot of "winby," follow the steps 1–6 given in Section 2.2, plus:

1. Click on **Stem-and-leaf** in the Descriptive box of the Explore: Plots dialog box.
2. Click on **Continue**.
3. Click on **OK** to run the procedure.

The stem-and-leaf plot will appear in the SPSS Viewer, as shown in Figure 2.12.

```
WINBY Stem-and-Leaf Plot

 Frequency      Stem &  Leaf

     9.00        0 . 111111111
    40.00        0 . 2222222233333333333333333333333333333333
    26.00        0 . 44444444444444445555555555
    36.00        0 . 666666666666667777777777777777777777
    11.00        0 . 88888899999
    29.00        1 . 00000000000000000000111111111
     4.00        1 . 2333
    27.00        1 . 444444444444444445555555555
    12.00        1 . 666666667777
    11.00        1 . 88888999999
     4.00        2 . 1111
     3.00        2 . 223
    14.00        2 . 44444455555555
     8.00        2 . 66677777
     5.00        2 . 88888
     3.00        3 . 111
     2.00        3 . 22
     6.00 Extremes    (>=34)

Stem width:        10
Each leaf:     1 case(s)
```

Figure 2.12 Stem-and-Leaf Display of Points by Which Games Were Won

The stem-and-leaf plot is similar to a histogram. In this case, the stem represents the 10's digit, and the leaf represents the 1's digit. The points displayed on the plot range from 1 [(0 × 10) + 1] to 32 [(3 × 10) + 2] points. The frequency column gives the number of observations in each interval. Notice that there are five sections of stem with each of the values 0, 1, and 2. The frequency column shows that 9 games were won by 1 point, 40 games were won by 2 or 3 points, and so on. Six of the games were won by 34 or more points, and are considered "extremes."

Chapter Exercises

2.1 Open the "bodytemp.sav" data file containing gender, body temperature, and pulse rate of 130 adults. These data were collected in part to examine whether normal body temperature is 98.6° Fahrenheit. Use the data set to conduct to following analyses and answer the following questions:

    a. Create a frequency distribution of body temperature. How many adults in the sample have a normal body temperature of 98.6° Fahrenheit?

    b. What percent of adults have a temperature less than 98.6° Fahrenheit?

    c. What is the lowest and highest temperature?

    d. Create a histogram of the temperature; adjust the graph so that there are 12 intervals.

2.2 Open the "fire.sav" data file, which contains demographic and performance data on 28 firefighter candidates, and use SPSS to answer the following:

    a. How many male firefighter applicants are there in the sample? What percentage of the total number of applicants is this?

    b. What percent of applicants are members of a minority group?

    c. For females only, what percentage of females had a time of less than 18 seconds on stair climb task? (Hint: you need to use the Select If command detailed in Chapter 1.)

    d. Repeat part c. for males only.

    e. Create a stem-and-leaf plot for the written test score.

    f. How many applicants had a score between 85 and 89 on the written test?

2.3 Open the 'titanic.sav" data file, which contains data on 2201 passengers on the Titanic. The variables are: gender, age category, class, and survival. Use SPSS to conduct the following analyses:

a. Create a bar chart of the class variable. Which class level had the most passengers?

b. Were there more first-class or second-class passengers?

c. How many passengers survived?

Chapter 3

Summarizing Data Numerically: Measures of Central Tendency

In addition to graphical summaries (Chapter 2), the primary features of a data set can be summarized through numerical indices. Measures of central tendency or location specify the "center" of a set of measurements. This chapter describes ways to use SPSS to obtain three common measures of location — the mode, the median, and the mean – of a sample. Measures of central tendency can be used to:

- find the most common college major for a group of students;
- find the midpoint of a set of ordered body weights that divides the set in half;
- calculate the average gross of the top movies from a given year;
- find the percentage of 13-year-old children who have a home computer.

3.1 THE MODE

The mode, especially useful in summarizing categorical or discrete numerical variables, is the category or value that occurs with the greatest frequency. One way to obtain the mode with SPSS for Windows is by using the Frequencies procedure. This is the same procedure used to obtain frequency distributions, histograms, and bar charts discussed in Chapter 2.

We illustrate how to obtain the mode using the "movies.sav" data file, which contains information on top grossing movies of 2001. Data include the genre, opening week gross, total gross, number of theatres in which it was released, studio, and the number of weeks the movie was in the top 60. The "genre" variable is a categorical variable representing the type of movie. To obtain the mode of this variable:

1. Click on **Analyze** from the menu bar.

2. Click on **Descriptive Statistics** from the pull-down menu.

3. Click on **Frequencies** from the pull-down menu.

4. Click on the "genre" variable and then the **right arrow button** to move the variable into the Variable(s) box.

5. Click on the **Statistics button** at the bottom of the screen. This opens the Frequencies: Statistics dialog box, as shown in Figure 3.1.

6. Click on the **Mode** option in the Central Tendency section.

7. Click on **Continue** to close this dialog box.

8. Click on **OK** to close the Frequencies dialog box and execute the procedure.

The output is shown in Figure 3.2.

Notice that in addition to the frequency distribution, the output lists the mode of the variable in the statistics table; it is genre "4." Because genre is a categorical variable, a value of 4 is representative of a particular genre. In this case, it represents "comedy." So, there are more comedies in the top 100 movies than any other genre.

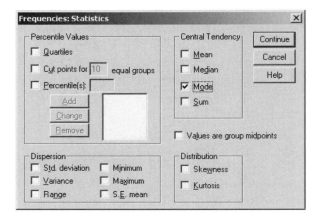

Figure 3.1 Frequencies: Statistics Dialog Box

Statistics

Movie type

N	Valid	100
	Missing	0
Mode		4

Movie type

		Frequency	Percent	Valid Percent	Cumulative Percent
Valid	Thriller/horror	11	11.0	11.0	11.0
	family	7	7.0	7.0	18.0
	drama	22	22.0	22.0	40.0
	comedy	38	38.0	38.0	78.0
	adventure/fantasy	22	22.0	22.0	100.0
	Total	100	100.0	100.0	

Figure 3.2 Frequency Distribution with Mode for Movie Genre

It is also possible to determine the mode of a variable by examining the frequency distribution itself. As an example, refer back to Figure 3.2. Even without the Mode option, you could search through the Frequency column for the row with the highest number, here comedy. Or, you could look at the Valid Percent column for the largest percentage, here 38.0%. The value associated with these numbers is the most common value for the variable — the mode of the variable.

3.2 THE MEDIAN AND OTHER PERCENTILES

The Median

The median is a value that divides the set of ordered (from smallest to largest) observations in half. That is, one-half the observations are less than (or equal to) the median value, and one-half the observations are greater than (or equal to) the median value. The symbol for the median is M.

The procedure for determining the median of a variable is similar to that for obtaining the mode. You simply need to click on the **Median** option instead of the Mode option in the Central Tendency box of the Frequencies: Statistics dialog box. (See Fig. 3.1.)

Continuing with the "movies.sav" data file we used in Section 3.1, we will examine the number of weeks the movies were in the Top 60. Your output should look like that in Figure 3.3. The median of the distribution is 14 weeks, as indicated on the first table in Figure 3.3.

Statistics

Weeks in top 60

N	Valid	100
	Missing	0
Median		14.00

Weeks in top 60

		Frequency	Percent	Valid Percent	Cumulative Percent
Valid	7	8	8.0	8.0	8.0
	8	6	6.0	6.0	14.0
	9	4	4.0	4.0	18.0
	10	4	4.0	4.0	22.0
	11	10	10.0	10.0	32.0
	12	6	6.0	6.0	38.0
	13	10	10.0	10.0	48.0
	14	5	5.0	5.0	53.0
	15	9	9.0	9.0	62.0
	16	5	5.0	5.0	67.0
	17	2	2.0	2.0	69.0
	18	6	6.0	6.0	75.0
	19	3	3.0	3.0	78.0
	20	2	2.0	2.0	80.0
	21	1	1.0	1.0	81.0
	22	2	2.0	2.0	83.0
	23	2	2.0	2.0	85.0
	24	6	6.0	6.0	91.0
	25	2	2.0	2.0	93.0
	26	1	1.0	1.0	94.0
	27	3	3.0	3.0	97.0
	29	1	1.0	1.0	98.0
	31	1	1.0	1.0	99.0
	38	1	1.0	1.0	100.0
	Total	100	100.0	100.0	

Figure 3.3 Frequency Distribution with Median for Total Gross of Movies

As with the mode, you can also determine the median from the frequency distribution. To do so, recall that the cumulative percentage column represents the percentage of cases at or below a given value. Because the median is the value of the ordered distribution at which 50% of the values are below it, to find the median, locate the first row in the distribution that has a cumulative percentage of 50 or greater. If it is not exactly 50%, the value associated with the percentage is the median. In this example, it is 53%. The value (number of weeks) associated with 53% is 14 weeks.

If the percentage in the cumulative percent column is exactly 50%, the median is halfway between that value and the subsequent value on the frequency distribution. This can occur when the distribution has an even number of observation. Then, median is halfway between: the observation at the $n/2$ position, and the observation at the $n/2 + 1$ position. For example, when there are 100 observations in the data set, the median is halfway between the 50th and 51st positions.

There are some additional features of the Frequencies Procedure that may be useful in some cases. For instance, if you would like to obtain both the mode and the median of a variable, you can select more than one option from the Frequencies: Statistics dialog box and obtain several statistics at once. There may also be times that you wish to only obtain the statistics, but not the frequency distribution. This is particularly useful for examining continuous variables from very large data sets. Suppose, for instance, you have a data file containing heights of 500 people. If the heights were measured to the nearest 100th of an inch, there would be very few data points with more than one observation. Thus, the frequency distribution would be a long ordered listing of the data points. There is an option on the Frequencies dialog box called "Display frequency tables" that governs whether or not the frequency distribution is displayed. The default for this option is "yes," but you may manually turn off the option by clicking on the box to the left of the phrase.

Quartiles, Deciles, Percentiles, and Other Quantiles

Just as the median divides the set of ordered observations into halves, quartiles divide the set into quarters, deciles divide the set of ordered observations into tenths, and percentiles divide the set of observations into hundredths. Quantile is a general term that includes quartiles, deciles, and percentiles.

You can obtain quartiles or percentiles from the Frequencies procedure by selecting the appropriate option in the upper left box of the Frequencies: Statistics dialog box. For instance, click on **quartiles** to generate a list of the quartiles. Deciles can be obtained by clicking on the **"Cut points"** option and selecting 10 equal groups. Other percentiles can be obtained by clicking on the **percentiles** option and entering the desired figures.

Statistics

Weeks in top 60

N	Valid	100
	Missing	0
Percentiles	25	11.00
	50	14.00
	75	18.75

Figure 3.4 Quartiles for Weeks in Top 60

Quartiles for the variable "weekstop" in the "movies.sav" data file are displayed in Figure 3.4. The 25th percentile is equivalent to the first quartile. Thus, one-quarter (25%) of the movies were in the top 60 for 11 or fewer weeks. The 50th percentile is the same as the median — 14 weeks. The 7^{5th} percentile, or third quartile is 18.75 weeks.

3.3   THE MEAN

The mean of a set of numbers is the arithmetic average of those numbers. The mean summarizes all of the units in every observed value, and is the most frequently used measure of central tendency for numerical variables. (When data are skewed, however, the median is generally a more appropriate measure of central tendency.) The symbol for the mean in a sample is x, which is often referred to as "x bar."

There are several methods for obtaining the mean of a distribution with SPSS for Windows. You can use the Frequencies procedure by clicking on **Mean** in the Central Tendency box in the Frequencies: Statistics dialog box. Try this with the movies data and the "weekstop" variable. The mean should be 15.26 weeks. In this example, the mean is somewhat larger than the median (14 weeks), suggesting that the distribution may be positively skewed. (In normally distributed distributions, the mean and median are similar in value.)

The mean can also be calculated with SPSS using the Descriptives or the Explore procedures. To obtain the mean using the Descriptives procedure:

1. Click on **Analyze** from the menu bar.
2. Click on **Descriptive Statistics** from the pull-down menu.
3. Click on **Descriptives** from the pull-down menu. This opens the Descriptives dialog box, as shown in Figure 3.5.
4. Move the "weekstop" variable to the Variable(s) box by clicking on the variable and then on the **right arrow button**.

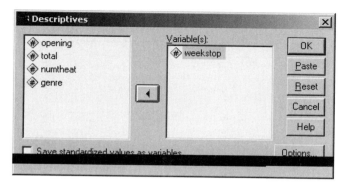

Figure 3.5 Descriptives Dialog Box

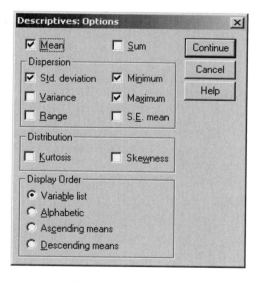

Figure 3.6 Descriptives: Options Dialog Box

5. Click on the **Options** button to open the Descriptives: Options dialog box (Fig. 3.6). The default options selected are: Mean, Std. deviation, Minimum and Maximum. We will accept the default in this example.

6. Click on **Continue** to close this dialog box.

7. Click on **OK** to run the procedure.

Did you obtain the same mean as you did when you used the Frequencies procedure?

You may also obtain the mean and several other descriptive statistics using the Explore procedure as follows:

Descriptives

			Statistic	Std. Error
Weeks in top 60	Mean		15.26	.632
	95% Confidence Interval for Mean	Lower Bound	14.01	
		Upper Bound	16.51	
	5% Trimmed Mean		14.88	
	Median		14.00	
	Variance		39.891	
	Std. Deviation		6.316	
	Minimum		7	
	Maximum		38	
	Range		31	
	Interquartile Range		7.75	
	Skewness		.939	.241
	Kurtosis		.747	.478

Figure 3.7 Output from the Explore Procedure

1. Click on **Analyze** from the menu bar.
2. Click on **Descriptive Statistics** from the pull-down menu.
3. Click on **Explore** from the pull-down menu.
4. Click on and move the "weekstop" variable to the Dependent List box using the **right arrow button**.
5. In the display box, click on the **Statistics** option.
6. Click on **OK**.

In addition to the mean, this procedure lists the median and several other descriptive statistics (see Fig. 3.7). For instance, the median is 14 weeks. We will discuss some of the other statistics, such as the variance, standard deviation, and range in the Chapter 4.

Proportion as a Mean

There is an exception to the rule that requires a variable to have numerical properties in order to calculate the mean — it is the dichotomous categorical variable. A dichotomous variable is a variable with only two possible values. If such a variable is coded with values 0 and 1, the mean will be the proportion of the cases with a value of 1. We will illustrate this using the "titanic.sav" data file as described in Chapter 2. The variable "survived' is coded so 0 = no, and 1 = yes. Thus, the mean represents proportion of passengers who survived the ship's sinking.

Statistics

SURVIVED

N	Valid	2201
	Missing	0
Mean		.32

SURVIVED

		Frequency	Percent	Valid Percent	Cumulative Percent
Valid	no	1490	67.7	67.7	67.7
	yes	711	32.3	32.3	100.0
	Total	2201	100.0	100.0	

Figure 3.8 Frequency Distribution and Mean of a Dichotomous Variable

Compute the mean using the Frequencies procedure. From the frequency distribution (Fig. 3.8), note that the percentage of survivors (coded 1) in the sample is about 32.3%. Within rounding, this is the same as the mean of .32.

Chapter Exercises

3.1 The file "library.sav" contains information on the size (number of books in the collection) and staff in 22 college libraries. Use this file to perform the following analyses with SPSS:

    a. Determine the mean and median of the distribution of staff using the Frequencies procedure. How do these measures compare? What do you think accounts for the differences?

    b. What value is at the 10th percentile? The 90th percentile?

    c. Create a histogram of the variable. Describe the distribution. Do any of the observations seem to be outliers? If so, how do these observations affect your findings in parts (a) and (b)?

3.2 Use the "fire.sav" data file, which contains demographic and test performance data on 28 firefighter applicants, to do the following:

    a. Determine the mean obstacle course time of the sample using either the Frequencies or the Descriptives procedure.

    b. Suppose you discovered that the clock device that kept time was set to start at 2 seconds, rather than 0 seconds. In order to obtain more accurate time, subtract 2 seconds from each candidate's time, and calculate the mean of the revised times. (Hint: use the Compute procedure.)

c. How does your result in (a) compare to that in part (b)? What principle does this illustrate?

d. Repeat the procedure, this time dividing each time by 2. How does the mean of the revised times compare to the mean of the original times?

3.3 The "IQ.sav" data file contains information on language and non-language IQ scores for a sample of children. Using these data use SPSS to complete the following:

a. Compute the mean, median, and mode of each type of IQ using the Frequencies procedure.

b. Which is the "best" measure of central tendency for the language IQ scores? Why? For the nonlanguage scores? Why?

3.4 The "movies.sav" data file contains information on top grossing movies of 2001. Following the steps below, use SPSS to illustrate the principle that the sum of all deviations from the mean is zero.

a. In Section 3.3 we found that the mean number of weeks these movies spent in the top 60 was 15.26 weeks. Compute a new variable, called "dev," representing deviations from this mean. Details for computing variables are contained in Chapter 1. The algebraic expression, which you will enter in the Numeric Expression box of the Compute Variable dialog box, is: "weekstop-15.26."

b. Now, compute the sum and mean of this new variable, "dev," using the Descriptives procedure.

c. Did you find that both the sum and mean are 0? Why is this?

Chapter 4

Summarizing Data Numerically: Measures of Variability

Chapter 3 described the use of SPSS to obtain measures of central tendency for a data set. Another important characteristic of numerical data is variability. Measures of variability indicate how spread out the observations are, that is, how much the values differ from individual to individual. For example, measures of variability let you:

- find the difference between the largest and smallest salary paid to people working at a particular company;

- determine how daily hours of sleep vary among different species of mammals;

- compare two or more distributions, such as whether basketball players or football players vary more in terms of height.

This chapter demonstrates ways to use SPSS to examine variability in a sample. We demonstrate how to calculate range, interquartile range, standard deviation, variance, and standard scores. A graphical display of measures of variability is demonstrated through box-and-whisker plots.

4.1   RANGES

The Range

The range is the difference between the maximum value in a distribution and the minimum value. There are several ways to find the range. One method is to create a frequency distribution of the scores as demonstrated in Section 2.2. To find the range from this distribution, subtract the minimum value from the maximum value. When you have a small data set with relatively few values, this method is adequate. However, if your data set has many different values, the frequency distribution will be very large and it is more efficient to compute the range using the Statistics option within the Frequencies procedure. First follow steps 1–5 given in Section 2.1 for creating a frequency distribution. Next:

1. Click on **Statistics** to open the Frequencies: Statistics dialog box (see Fig. 2.2).
2. Click on **Range** in the Dispersion box.
3. Click on **Continue**.
4. Click on **OK**.

A second method for obtaining the range and other measures of dispersion uses the Descriptives procedure. We shall illustrate with the "sleep.sav" data file, which contains information on physical, environmental, and sleep characteristics of 62 mammals. Let us find the range of hours of sleep per day. After opening the file:

1. Click on **Analyze** on the menu bar.
2. Click on **Descriptive Statistics** from the pull-down menu.
3. Click on **Descriptives** to open the Descriptives dialog box (see Chapter 3, Fig. 3.5).
4. Click on the variable name for which you wish to have the range ("totsleep"), and then click on the **right arrow button** to move the variable into the Variables box.
5. Click on **Options** to open the Descriptives: Options dialog box (Fig. 3.6).
6. In the Dispersion box, click on **Range**. (Notice that the mean, standard deviation, minimum, and maximum boxes are already checked. This is the default.)
7. Click on **Continue**.
8. Click on **OK**.

The summary statistics for the Descriptives procedure should appear as shown in Figure 4.1.

Descriptive Statistics

	N	Range	Minimum	Maximum	Mean	Std. Deviation
TOTSLEEP	58	17.30	2.60	19.90	10.5328	4.60676
Valid N (listwise)	58					

Figure 4.1 Descriptives for Hours of Sleep Variable

The variable name appears in the left column of the table, followed by the descriptive statistics. The range, the difference between the maximum (19.90 hours) and the minimum (2.60 hours), is 17.30 hours. Notice that the N is 58. Thus, four of the mammals (62 mammals are in the data file) did not have data on total hours of sleep.

The Interquartile Range

Quartiles are one of the measures of location discussed in Chapter 3. The interquartile range is the difference between the first and third quartiles. The interquartile range is a more stable measure of variability than the range because it is less affected by extreme scores. The interquartile range can be obtained by using the Explore procedure, as described in Section 3.3. The interquartile range for total sleep is 5.575 hours.

4.2   THE STANDARD DEVIATION

The standard deviation (s) is a type of average of the distances of the values of individual observations from the mean. It is one of the most common and most useful measures of dispersion.

The variance (s^2) is the square of the standard deviation. If x_i ($i = 1,2,...,n$) is the value of observation i, and $\overline{x}$ is the sample mean, then the sample variance is:

$$s^2 = \frac{\sum (x_i - \overline{x})^2}{n - 1}$$

The standard deviation is found by computing the square root of the variance, s = $\sqrt{s^2}$. The Descriptives procedure is one method for producing the standard deviation in SPSS.

Descriptive Statistics

	N	Minimum	Maximum	Mean	Std. Deviation	Variance
TOTSLEEP	58	2.60	19.90	10.5328	4.60676	21.222
Valid N (listwise)	58					

Figure 4.2 Descriptives for Hours of Sleep Variable Including the Variance

1. Click on **Analyze** on the menu bar.
2. Click on **Descriptive Statistics** from the pull-down menu.
3. Click on **Descriptives** to open the Descriptives dialog box.
4. Click on the variable name that you wish to examine and the **right arrow button** to move it to the Variable(s) box.

By default, the output will contain the mean, standard deviation, minimum, and maximum values. To obtain the variance, follow steps 1--4 above for the standard deviation, and then:

1. Click on **Options**.
2. Click on **Variance** in the Dispersion box.
3. Click on **Continue** and then click on **OK**.

The output is shown in Figure 4.2. Notice that the variance is the square of the standard deviation; the variance is $4.60676^2 = 21.222$.

4.3   SOME USES OF LOCATION AND DISPERSION MEASURES TOGETHER

Box-and-Whisker Plots

A box-and-whisker plot is a useful graphical display that shows the median, interquartile range, and extremes of a data set. SPSS for Windows creates these plots using the Explore procedure. To illustrate this procedure, we continue with the "sleep.sav" data file, but this time with the "lifespan" variable, which indicates the maximum life span (in years) for the mammals. To create the box-and-whisker plot:

1. Click on **Analyze** on the menu bar.
2. Click on **Descriptive Statistics** from the pull-down menu.
3. Click on **Explore** to open the Explore dialog box.
4. Click on the "lifespan" variable then click the **top right arrow button**.

5. Click on the **Plots button** to open the Explore: Plots dialog box.
6. Click off the **stem-and-leaf** option in the descriptive section and leave clicked the default **Factor levels together** in the boxplots section.
7. Click on **Continue**.
8. Click on **OK**.

The output is shown in Figure 4.3; we are primarily concerned with the graph. The red box in the box-and-whisker plot represents the interquartile range. The top of the red box demarcates the third quartile (28 years), and the bottom denotes the first quartile (6.4 years). (You may wish to check these figures by using the Frequencies procedure to report the quartiles.) The horizontal line in the middle of the red box represents the median of the distribution. Here, it is 15.1 years.

The horizontal line (whisker) at 2 years indicates the minimum life span. The whisker at 50 years denotes the maximum lifespan, <u>excluding outliers</u>. The circle and asterisk at 69 and 100 years, respectively, represent outliers.

Standard Scores

Standard scores, also called z-scores, indicate the relative position of a single observation in the sample, that is, the number of standard deviations the observation is above or below the mean. For values above the mean, the z-score is positive. Values below the mean have negative z-scores, and values equal to the mean have a z-score of 0. The z-scores are calculated using the Descriptives procedure.

We shall illustrate using the "sleep.sav" data file. To create standard scores for the variable "lifespan":

1. Click on **Analyze** on the menu bar.
2. Click on **Descriptive Statistics** and then **Descriptives** to open the Descriptives dialog box.
3. Click on the "lifespan" variable and move it into the Variable(s) box with the **right arrow button**.
4. Click on the **Save standardized values as variables** box.
5. Click on **OK**.

This will cause SPSS to create a new z-score variable. By default, the new variable is named by prefixing the letter Z to the original variable. For example, "lifespan" becomes "Zlifespan." You can examine the standard scores using the Frequencies procedure (see Section 3.1). The histogram is displayed in Figure 4.4.

Descriptives

			Statistic	Std. Error
LIFESPAN	Mean		19.8776	2.39060
	95% Confidence Interval for Mean	Lower Bound	15.0905	
		Upper Bound	24.6647	
	5% Trimmed Mean		17.8594	
	Median		15.1000	
	Variance		331.468	
	Std. Deviation		18.20626	
	Minimum		2.00	
	Maximum		100.00	
	Range		98.00	
	Interquartile Range		21.6250	
	Skewness		2.014	.314
	Kurtosis		5.885	.618

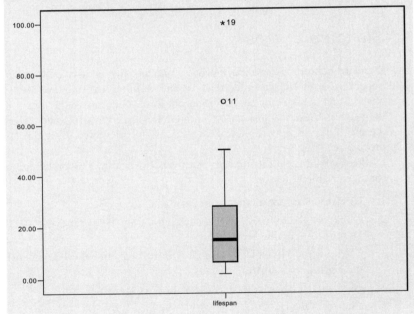

Figure 4.3 Box-and-Whisker Plot of Maximum Lifespan of Mammals

Statistics

Zscore(LIFESPAN)

N	Valid	58
	Missing	4
Mean		.0000000
Std. Deviation		1.000000
Minimum		-.98195
Maximum		4.40082

Histogram

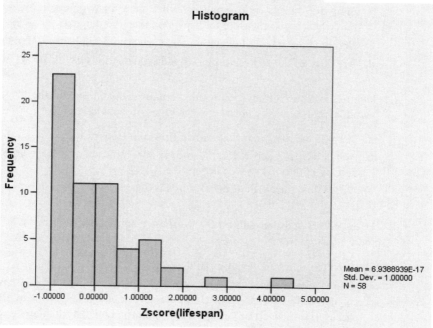

Mean = 6.9388939E-17
Std. Dev. = 1.00000
N = 58

Figure 4.4 Output from Frequencies Procedure Summarizing "Zlifespan"

By definition, the mean and standard deviation of the z-scores are 0 and 1, respectively. If the distribution of "lifespan" were normal, than the range of "Zlifespan" would be generally between –2.0 and 2.0. But, as we discovered with the box-and-whisker plot (and can see again in the histogram in Figure 4.4), this variable has a positively skewed distribution. The two outliers have standardized scores greater than 2.0.

Chapter Exercises

4.1 The "library.sav" data file contains measurements staff and size of 22 college libraries. Use this data file to perform the following analyses with SPSS for Windows:

    a. Find the range of the number of volumes held by the libraries using either the Frequencies and Descriptives procedures.

    b. Find the interquartile range. In what situation would the interquartile range be a better measure of dispersion than the range?

    c. Compute the standard deviation and variance of number of volumes.

    d. Are most of the scores within 2 standard deviations of the mean?

4.2 Using the "movies.sav" data file, compute z-scores for the "opening" variable, represents the opening week gross of the movies.

    a. What is the z-score for the movie *Cats and Dogs*?

    b. What is the z-score for the movie that grossed the least amount of money in the first week? The most?

    c. Verify that the mean of the z-scores is 0.

4.3 Using SPSS and the data on 28 firefighter applicants in "fire.sav" file, do the following:

    a. Find the standard deviation of the candidates' body drag time.

    b. Suppose that each firefighter received agility training and decreased his or her time by 1 second. Compute the standard deviation of the new times (Hint: use the Compute command). Did the standard deviation change? Why or why not?

    c. Suppose that each firefighter received more training and halved his or her original time. Compute the new scores and find the standard deviation of them. How did dividing by a constant affect the standard deviation?

4.4 Use the "hotdog.sav" data set containing information on number of calories and milligrams of sodium in three types of hot dogs (beef, meat, and poultry) to do the following:

    a. Use the Explore procedure to produce summary statistics and a box-and-whisker plot for calories each of the three types of hotdog. (HINT: the "type" variable will be used in the Factor List box and "calories" in the Dependent List box.)

    b. What is the median number of calories for each type of hot dog?

    c. What is the minimum and maximum number of calories for beef hot dogs?

    d. Are there any outliers for the poultry hot dogs?

    e. Based on the standard deviations, which type of hot dog has the most variability in calories?

Chapter 5

Summarizing Multivariate Data: Association Between Numerical Variables

In this chapter, we illustrate methods for summarizing the relationship or association between two or more variables measured on numerical scales. (Chapter 6 discusses association among categorical variables.) This association can be expressed either graphically or numerically. The graphical technique is the scatter plot, and the numerical index is the correlation coefficient. In the sample, the correlation coefficient is represented by r.

These methods can be used to answer questions about the relationship between variables such as:

- Is there a relationship between scores on the language and nonlanguage portions of an IQ test? That is, do students with high language scores also have high nonlanguage scores? (If so, this would be a positive association.)

- Is there a relationship between the number of days students are absent from class and their score on the final exam? That is, do students with a large number of absences have lower grades, whereas those with few absences have higher grades?

- Is the degree of exposure to radioactive materials associated with the rate of cancer mortality?

- Is there a relationship between fat content and number of calories in different breakfast cereals?

This chapter describes how to obtain scatter plots and correlation coefficients between numerical variables using SPSS for Windows.

5.1   ASSOCIATION OF TWO NUMERICAL VARIABLES

Scatter Plots

A scatter plot is a graphical technique used to illustrate the association of two numerical variables. Data are represented visually by making a graph with two axes: horizontal (x axis) and vertical (y axis). Each point in the plot represents one observation. When all observations are plotted, the diagram conveys information about the direction and magnitude of the association of the two variables (x and y).

To illustrate the use of SPSS for scatter plots, let us examine the relationship between language and nonlanguage IQ for 23 second-grade children. The data are contained in the file "IQ.sav."

To obtain a scatter plot, do the following:

1. Click on **Graphs** from the menu bar.
2. Click on **Scatter/Dot** from the pull-down menu.
3. Click on **Simple Scatter** and then on **Define** to open the Simple Scatterplot dialog box (Fig. 5.1).
4. Click on the "nonlanguage IQ" variable and move it to Y Axis box.
5. Click on the "language IQ" variable and move it to X Axis box.
6. Click on **OK** to close this dialog box and create the scatter plot.

The SPSS Viewer contains the scatter plot like that in Figure 5.2. Each point on the plot represents one observation. For example, one person had a language IQ score of 84 and a nonlanguage IQ score of 30, as marked in the graph. Another individual had scores of 109 and 74, respectively. Can you find it on the graph?

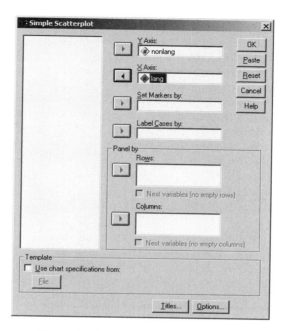

Figure 5.1 Simple Scatter Plot Dialog Box

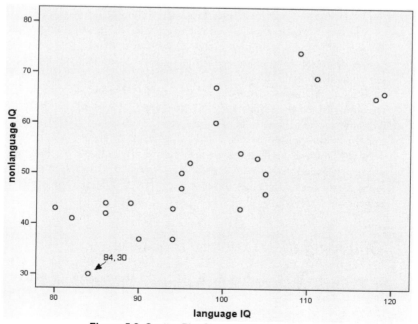

Figure 5.2 Scatter Plot Showing Positive Association

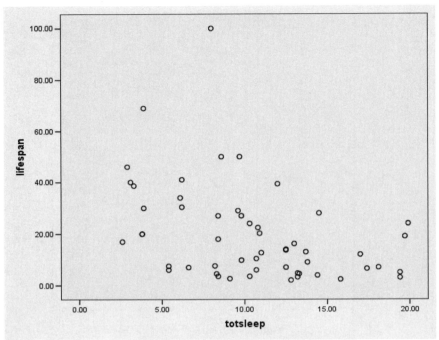

Figure 5.3 Scatter Plot Showing Negative Association

There is a positive association between language IQ and nonlanguage IQ. That is, individuals with high language IQ scores also tend to have high nonlanguage IQ scores, and those with low scores on one measure tend to have low scores on the other.

We will create another scatter plot using the data in "sleep.sav." In Chapter 4, we summarized the "lifespan" and "totsleep" variables separately. Now let us examine the relationship between the two by creating a scatter plot with total hours of sleep per day on the x-axis and lifespan on the y-axis. Following steps 1–6 will produce the results shown in Figure 5.3. In this plot, there appears to be a negative relationship between the two variables. That is, animal species that sleep many hours per day tend to have a lower life span than those who sleep fewer hours per day.

Changing the Scales of the Axes

SPSS chooses the scales for the *x* and *y* axes that best fit the range of the data, but you may manually adjust the scales if you wish. Below are steps to edit the *x*-axis scale of any scatter plot:

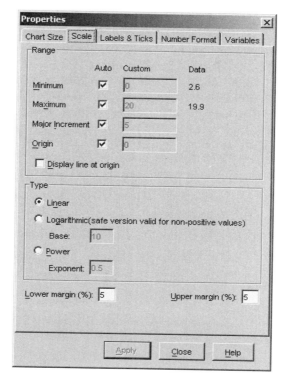

Figure 5.4 Scale Tab of Properties Dialog Box for X Axis of the Total Sleep by Life Span Scatter plot

1. Double click on the chart in the SPSS Viewer. This opens the Chart Editor.
2. Select **Edit** from the menu bar.
3. Click on **Select X Axis** from the pull-down menu to open the Properties Dialog box.
4. Select the **Scale** tab (Fig. 5.4).
5. Edit the Minimum and Maximum points of the range by clicking off the **Auto** selections and entering the desired numbers in the Custom sections.
6. Click **Apply** to redraw the scatter plot and then **Close** to apply the changes and close the Properties dialog box.
7. Click the **X** in the top right corner to close the Chart Edit Window.

Other Information Revealed by Scatter Plots

Examining a scatter plot can reveal important information. As we have discussed, it is possible to discern positive and negative associations. Scatter plots lso allow one to determine, for instance, when a relationship between two variables is nonlinear and/or when bivariate outliers exist. We will illustrate the latter of these cases.

The data file "IQ.sav" contains information regarding language and nonlanguage IQ scores for 23 students. The range of the language scores is 80--119, and the range of the nonlanguage scores is 30--74. As we saw in Figure 5.2, the scatter plot of these two variables shows a positive association.

Now, open the data file "IQ2.sav." This file contains the same scores in the original "IQ.sav" file, plus one additional data point, a student with a language score of 80 and a nonlanguage score of 72. Considering each of these scores alone, neither is an outlier; each is within the range of the original scores for its variable.

Create the scatter plot for the variables with language on the x-axis and nonlanguage on the y-axis (see Fig. 5.5). Notice the outlier; we have labeled the point (80,72). This point represents an individual with a very low language IQ and a very high nonlanguage IQ. There are no other data points in its vicinity, and it is clearly an outlier. The data analyst should attempt to understand why this unusual pairing of values occurred.

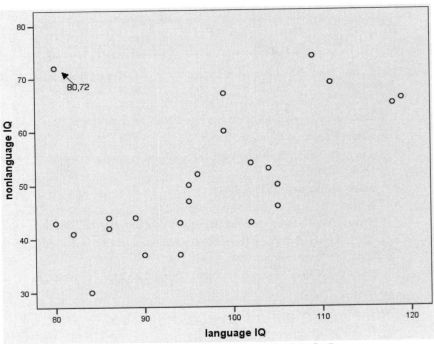

Figure 5.5 Scatter Plot Showing a Bivariate Outlier

The Correlation Coefficient

The Pearson correlation coefficient indicates the degree of linear association of two numerical variables. The correlation coefficient ranges from −1 to +1. A positive value (e.g., .10, .40, .80) reflects a direct associations between the two variables and a negative value (e.g., −.20, −.40, −.80) reflects a negative or inverse relationship. The strength of association is indicated by the absolute value of the correlations; for example, the values −.80 and .80 represent equally strong relationships. Zero is the weakest correlation, and 1 (or −1) the strongest. As a rule of thumb, correlations between 0 and .30 (absolute value) are considered weak; those between .31 and .60 (absolute value) are considered moderate, and those greater than .60 (absolute value) are considered strong.

We will illustrate the procedure for calculating the correlation coefficient using the "cereal.sav" data file. This file contains nutritional information on 77 brands of cereal. We will examine the relationship between sugar content and calories. You may begin by opening the file. To compute the correlation coefficient for these two measured variables:

1. Click on **Analyze** from the menu bar.
2. Click on **Correlate** from the pull-down menu.
3. Click on **Bivariate** from the pull-down menu. This opens the Bivariate Correlations dialog box (see Fig. 5.6).
4. Click on the "sugar" and "calories" variables and move them to the Variables box by clicking on the **right arrow button**.
5. Click on **OK** to run the procedure.

The output should look like that in Figure 5.7.

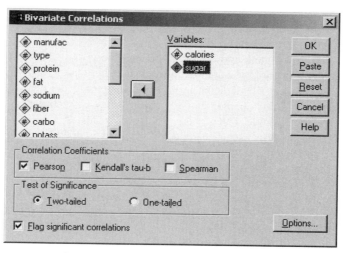

Figure 5.6 Bivariate Correlations Dialog Box

Correlations

		calories	sugar
calories	Pearson Correlation	1	.562**
	Sig. (2-tailed)		.000
	N	77	77
sugar	Pearson Correlation	.562**	1
	Sig. (2-tailed)	.000	
	N	77	77

· Correlation is significant at the 0.01 level

Figure 5.7 Correlation of Sugar Content and Calories

SPSS lists the correlation coefficients it calculates in a correlation matrix. The values on the diagonal are all 1 because they represent the correlations of each variable with itself. The values above and below the diagonal are identical. In the example, the correlation between amount of sugar and calories is .562. This is a moderate, positive association; cereals with higher sugar content tend to have more calories.

Let us repeat this procedure two more times, first with the "IQ.sav" data file, and then with the "IQ2.sav" data file. For the original data file, you should determine that the correlation is .769. Simply by adding one outlying observation ("IQ2.sav"), the correlation coefficient decreases to .561. This is a dramatic change, and underscores the need to examine the scatter plot of two variables in addition to calculating the coefficient.

Rank Correlation

A (Spearman) rank correlation is the correlation coefficient computed from two variables that are measured on an ordinal rather than a numerical scale. Because the coefficient is calculated in a similar manner to the Pearson coefficient, it is discussed in this chapter. Some variables may be recorded as ranks in the first place, for example, fastest runner (1), second fastest (2), ..., slowest runner (n). Data on numerical scales can be re-expressed as ranks. For example, the student scoring 100 on a test may be given rank 1, the student with 96 given rank 2, and so on. Let us repeat the calculation of the correlation of sugar content and calories, this time using the Spearman procedure. Because these data are on a refined numerical scale, we shall rank them in order to illustrate the rank correlation procedure. (Note: we do this only for illustrative purposes; generally, when two variables are measured on a numerical scale, the Pearson correlation is a more precise measure of linear association.)

To rank these two variables:

1. Click on **Transform** from the menu bar.
2. Click on **Rank Cases** from the pull-down menu. This opens the Rank Cases dialog box (see Fig. 5.8).
3. Move the "sugar" and "calories" variables to the Variable(s) box by clicking on the name of each variable and then on the **top right arrow button**.
4. Click on **OK**.

SPSS creates two new variables, "rcalorie" and "rsugar," which consist of ranked data. (SPSS automatically creates the new variable names by adding an r-prefix to the original variable names.) The procedure to compute the Spearman correlation for these two variables is similar to the steps outlined for calculating the Pearson coefficient. The only difference is that you must click off the Pearson option (the default) and click on the Spearman option in the Correlation Coefficients box (see Fig. 5.6).

Your output should look like Figure 5.9.

The correlation of mortality and exposure, when ranked, is .596. This is slightly larger, but consistent with, the Pearson correlation coefficient, .562.

5.2   *MORE THAN TWO VARIABLES*

Correlation Matrix

In some instances, you may be interested in examining the correlation between many pairs of variables. SPSS can calculate many pairs of correlations at one time. The correlations for each pair of measures are computed and arranged in a matrix that has one row for each variable and one column for each variable.

The procedure is the same as that detailed in Section 5.1, but you need to include the names of all of the variables for which you desire correlations in the Variables box of the Bivariate Correlations dialog box.

To illustrate, open the "fire.sav" data file and create a correlation matrix for the following variables: "stair," "body," "obstacle," "agility," "written," and "composit" by including the names of all of them in the Variables box. Your output should look like Figure 5.10.

To find the correlation between body drag test time and obstacle course time, for instance, find the intersection of the column labeled body and the row labeled obstacle in Figure 5.10. The correlation coefficient is .759. (Note: this is the same number that is found at the intersection of the row labeled body and the column labeled obstacle. In other words, the correlation matrix is symmetrical.)

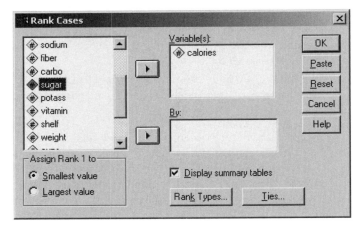

Figure 5.8 Rank Cases Dialog Box

Correlations

			RANK of CALORIES	RANK of SUGAR
Spearman's rho	RANK of CALORIES	Correlation Coefficient	1.000	.596**
		Sig. (2-tailed)	.	.000
		N	77	77
	RANK of SUGAR	Correlation Coefficient	.596**	1.000
		Sig. (2-tailed)	.000	.
		N	77	77

· Correlation is significant at the .01 level (2-tailed).

Figure 5.9 Spearman Correlation Output

It is also possible to discern patterns among correlations in the matrix. For example, the correlations among body drag, stair climb, and obstacle course times are all positive and moderately strong to strong (ranging from .734 for obstacle course with stair climb, to .906 for stair climb with body drag). Because all are measures of athletic behavior, the positive association is to be expected. The written test score is negatively correlated with these tasks. Thus, more agile applicants (those with lower time scores) tend to have higher scores on the written test, and vice versa.

Correlations

		stair	body	obstacle	agility	written	composit
stair	Pearson Correlation	1	.906**	.734**	.954**	-.337	-.898**
	Sig. (2-tailed)		.000	.000	.000	.079	.000
	N	28	28	28	28	28	28
body	Pearson Correlation	.906**	1	.759**	.962**	-.466*	-.940**
	Sig. (2-tailed)	.000		.000	.000	.012	.000
	N	28	28	28	28	28	28
obstacle	Pearson Correlation	.734**	.759**	1	.875**	-.495**	-.875**
	Sig. (2-tailed)	.000	.000		.000	.007	.000
	N	28	28	28	28	28	28
agility	Pearson Correlation	.954**	.962**	.875**	1	-.458*	-.970**
	Sig. (2-tailed)	.000	.000	.000		.014	.000
	N	28	28	28	28	28	28
written	Pearson Correlation	-.337	-.466*	-.495**	-.458*	1	.660**
	Sig. (2-tailed)	.079	.012	.007	.014		.000
	N	28	28	28	28	28	28
composit	Pearson Correlation	-.898**	-.940**	-.875**	-.970**	.660**	1
	Sig. (2-tailed)	.000	.000	.000	.000	.000	
	N	28	28	28	28	28	28

** Correlation is significant at the 0.01 level (2-tailed).
* Correlation is significant at the 0.05 level (2-tailed).

Figure 5.10 Correlation Matrix

Missing Values

Only those cases with values for both variables can be used in computing a correlation coefficient. There are two ways to cause SPSS to eliminate cases with missing values: "listwise deletion" and "pairwise deletion." (See Chapter 1 for a discussion of missing values.) As an example, suppose that the third case in our data file were missing a score on the written test. Because the third case does not have complete information on all of the variables, listwise deletion eliminates the third case from the computation of all correlation coefficients. All correlations are calculated from the remaining 27 cases. Pairwise deletion eliminates the third case when computing only those correlations that involve the written test; thus some coefficients would be based on 27 observations and others would be based on all 28.

The default option in SPSS is pairwise deletion, which uses the maximum number of cases for each coefficient. You may request listwise deletion be clicking on the **Options button** of the Bivariate Correlations dialog box (Fig. 5.6) and choosing **Exclude cases listwise**.

Chapter Exercises

5.1 In Section 5.1 of this chapter, we created the scatter plot of total hours of sleep with lifespan for the 62 mammals in the "sleep.sav" data file.

a. Refer to the graph (Fig. 5.3) and estimate the value of the correlation coefficient. Check your estimate by calculating the Pearson correlation using SPSS.

b. Eliminate any outliers and recalculate the correlation. How did this affect your results?

5.2 The "cancer.sav" data file contains data on cancer mortality and index of exposure to nuclear materials for residents in counties near a nuclear power plant. Using this data file:

a. Compute the Pearson correlation coefficient between "expose" and "mortalit" and comment on the strength and direction.

b. Rank the two variables using the Transform procedure, and compute the Spearman correlation coefficient of the ranked variables.

c. How do the two coefficients compare?

5.3 The data file "enroll.sav" contains information from a random sample of 26 school districts. Information was obtained on the following variables:

(1) district enrollment;

(2) the percentage of students in the district who are African-American;

(3) the percentage of students who pay full price for lunches;

(4) an index of racial disproportion in classes for emotionally disturbed children (which is positive if the proportion of African-American students is greater than the proportion of white students).

Using this data file, compute a correlation matrix for all four variables and use it to answer the following questions:

a. Which correlation is largest (in magnitude)?

b. Explain what is meant by the negative correlation between enrollment and percent of African-Americans.

c. Racial disproportion is most highly associated with which other variable? What is the magnitude and direction of the association?

5.4 It has been shown that the relationship between amount of stress and work productivity is curvilinear. In other words, extremely low and extremely high amounts of stress are related to low work productivity, but moderate amounts of stress are associated with the maximum amount of productivity.

a. Create a hypothetical data set with two variables — amount of stress and work productivity — that you think will illustrate this relationship. Your data file should have a minimum of 20 observations.

b. Use SPSS to produce a scatter plot of your data. Did you succeed in simulating a curvilinear relationship?

c. Name another instance in which you might find a curvilinear relationship between two variables.

Chapter 6

Summarizing Multivariate Data: Association Between Categorical Variables

Chapter 5 describes how to use SPSS to summarize associations between numerical variables. In this chapter, we describe how to summarize two or more categorical variables. These analyses can be used to answer questions such as:

- Is there a relationship between political party and gender? For example, do women tend to vote for Democrats and men for Republicans? Or vice versa?

- Is there a relationship between region of the country and attitude toward capital punishment? For instance, do people in the South tend to favor the death penalty whereas those in the Northeast tend to oppose it?

- Were there more female than male survivors on the Titanic?

6.1   TWO-BY-TWO FREQUENCY TABLES

The relationship between two or more categorical variables is summarized using frequency tables, or cross-classifications of observations. Two-by-two tables are created when you have two variables, each with two possible outcomes. For example, we may have a sample of people categorized by race (minority, nonminority) and by gender. Frequencies for either variable separately are obtained using the Frequencies procedure. However, it cannot be determined from these individual frequency distributions how many minority males are in the sample, for example. This is accomplished using the Cross-tabulation procedure, which examines the counts of simultaneous occurrences of several values.

Open the data in the football data file ("football.sav"), which has record of favored team (home or away team) and winning team (home or away team) for 250 NFL games. To create a two-way frequency table of favored team by winning team, follow these steps:

1. Click on **Analyze** from the menu bar.
2. Click on **Descriptive Statistics** from the pull-down menu.
3. Click on **Crosstabs** to open the Crosstabs dialog box shown in Figure 6.1.
4. Highlight the "favored" variable by clicking on it, and then move it to the Row(s) box by clicking on the **top right arrow button**.
5. Highlight the "winner" variable by clicking on it, and then move it to the Column(s) box by clicking on the **middle right arrow button**.
6. Click on **OK**.

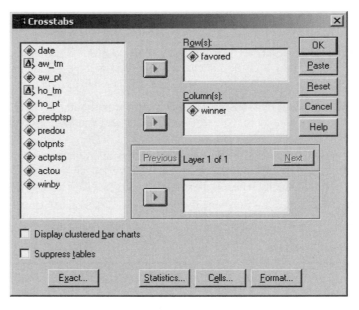

Figure 6.1 Crosstabs Dialog Box

Case Processing Summary

	Cases					
	Valid		Missing		Total	
	N	Percent	N	Percent	N	Percent
favored * winner	250	100.0%	0	.0%	250	100.0%

favored * winner Crosstabulation

Count

		winner		Total
		home	away	
favored	home	125	56	181
	away	31	38	69
Total		156	94	250

Figure 6.2 Cross-tabulation of Favored by Winner

This will produce a cross-tabulation of the row ("favored") by column ("winner") variables as shown in Figure 6.2. The number of cases for each combination of values of the row by column variables is displayed in the cells of the table.

The Case Processing Summary table indicates that there are no missing values for either variable. Examining the favored * winner Cross-tabulation table, we see that favored has two values (home and away) and winner has two values (home and away). A two-by-two table such as this has four "cells."

The number that appears in each of the cells is called the Count (or frequency); this is the number of cases in that cell. In this sample, there are 125 games in which the home team was the favored team <u>and</u> the winning team. In 56 games, the home team was favored and the away team won.

The rows and columns labeled "Total" are the marginal totals and represent the counts for the values of favored and winning team separately (the counts that would be obtained from a simple frequency distribution). For example, in the favored (row) marginals, we see that there were 181 games in which the home team was favored and 69 in which the away team was favored. Likewise, the 156 and 94 winner (column) totals indicate the number of games the home team and away team won, respectively. The number 250 represents the total number of games in the data file.

Calculation of Percentages

In addition to the count for each cell, SPSS will also calculate row, column, and total percentages. The row percentage is the percentage of cases in a row that fall into a particular cell. The column percentage is the percent of cases in a column that fall into a particular cell. The total percentage is the number of cases in a cell expressed as a percentage of the total number of cases in the table.

Use the file "titanic.sav" to illustrate these percentages. This data file contains sex and survival information for the 2201 passengers on the Titanic. We will run the Crosstabs procedure with "sex" as the row variable and "survived" as the column variable. This will address the question, was it really a case of "women first" — that is, were women evacuated to lifeboats ahead of men?

To calculate row, column, and total percentages, follow steps 1–5 in the preceding section, and then:

1. Click on **Cells** to open the Crosstabs: Cell Display dialog box.
2. Click on each box in the Percentages box to indicate that you wish you calculate **Row**, **Column**, and **Total** percentages.
3. Click on **Continue**.
4. Click on **OK**.

Your output should appear as in Figure 6.3.

The top number in each cell is the count. The cell counts indicate that there were 126 female nonsurvivors, 344 female survivors, 1364 male nonsurvivors, and 367 male survivors. The marginal counts indicate that there were 470 females and 1731 males, and that there were 1490 nonsurvivors and 711 survivors. The total number of passengers, 2201, appears in the bottom right-hand corner of the table.

SEX * SURVIVED Crosstabulation

| | | | SURVIVED | | |
			no	yes	Total
SEX	female	Count	126	344	470
		% within SEX	26.8%	73.2%	100.0%
		% within SURVIVED	8.5%	48.4%	21.4%
		% of Total	5.7%	15.6%	21.4%
	male	Count	1364	367	1731
		% within SEX	78.8%	21.2%	100.0%
		% within SURVIVED	91.5%	51.6%	78.6%
		% of Total	62.0%	16.7%	78.6%
Total		Count	1490	711	2201
		% within SEX	67.7%	32.3%	100.0%
		% within SURVIVED	100.0%	100.0%	100.0%
		% of Total	67.7%	32.3%	100.0%

Figure 6.3 Cross-tabulation of Sex by Survived with Percentages

Below the count is the "% within SEX," or the row percentage. These figures represent the percentages by values of sex (male and female). That is, 26.8% of females did not survive (which is calculated by dividing 126 by the total number of females in the sample (470) and multiplying by 100). Conversely, 78.8% of the males did not survive (1364   1731 × 100). Thus, it seems that women were given preferential treatment and were more likely to survive.

Figure 6.3 also displays "% within SURVIVED." For example, of the 1490 people who did not survive, 8.5% females (126   1490 × 100). The fourth line of each cell, labeled "% of Total," contains the total percentages. For example, the 344 female survivors represent 15.6% of the total 2201 passengers.

Phi Coefficient

The phi coefficient is an index of association for a two-by-two table. Like the correlation coefficient, its value is between −1 and 1. Values close to −1 or 1 indicate strong association of two variables, whereas values close to zero indicate a weak association. To compute the phi coefficient, follow steps 1–5 in Section 6.1, and then:

1. Click on **Statistics** to open the Crosstabs: Statistics dialog box.
2. Click on **Phi and Cramer's V** in the Nominal box.
3. Click on **Continue**.
4. Click on **OK**.

For example, the value of the phi coefficient for the data in Figure 6.3 is −.456. This indicates a moderate association of sex with survival; the majority of females survived, and the majority of males did not.

6.2   LARGER TWO-WAY FREQUENCY TABLES

In many studies, categorical variables have more than two values. The number of categories is not limited to two, and virtually any size row-by-column table is possible. For example, "political affiliation" may be recorded in three categories: Conservative, Liberal, and Moderate; "occupation" may have 15 categories representing 15 different job titles. The procedure for calculating counts and percentages, as well as the interpretations of the frequencies, is the same as described in Section 6.1.

As an example, open the "spit.sav" data file. This is a study on the effectiveness of two interventions to help major league baseball players stop using spit tobacco.[*] The frequency table for this example is a 3-x-2 table; there are two variables, "outcome" with three levels and "intervention" with two levels.

Using this data set, follow the steps in Section 6.1 to create a frequency table containing counts, and row, column, and total percentages. Your results, omitting the case-processing summary, should look like Figure 6.4. The row variable is the length of intervention (minimum or extended), and the column variable is player outcome (quit; tried to quit; failed to quit). Looking just at the total row, we see that 5 players successfully quit, 15 tried to quit but were unsuccessful, and 34 failed to quit. So, because overall only 9.3% of the players (5 divided by 54) were able to quit, the interventions were not particularly successful.

But what about the relationship between the two variables? That is, is one of the intervention types more successful than the other? To explore this, we look at the "percent within intervention group" figures. We note that of the players in the minimum intervention group, 0% quit successfully, compared with 19.2% in the extended intervention group. Similarly, we see that 89.3% of the minimum intervention group failed to quit, compared with 34.6% of the extended intervention group. Therefore, the extended intervention seems relatively more successful for this sample of baseball players.

intervention group * outcome of intervention Crosstabulation

| | | | outcome of intervention | | | |
			quit	tried	failed	Total
intervention group	minimum	Count	0	3	25	28
		% within intervention group	.0%	10.7%	89.3%	100.0%
		% within outcome of intervention	.0%	20.0%	73.5%	51.9%
	extended	Count	5	12	9	26
		% within intervention group	19.2%	46.2%	34.6%	100.0%
		% within outcome of intervention	100.0%	80.0%	26.5%	48.1%
Total		Count	5	15	34	54
		% within intervention group	9.3%	27.8%	63.0%	100.0%
		% within outcome of intervention	100.0%	100.0%	100.0%	100.0%

Figure 6.4 Cross-tabulation of Outcome by Intervention

---

[*]Data reproduced from summary tables in Greene, J.C., Walsh, M.M., & Masouredis, C. (1994). Report of a pilot study: A program to help major league baseball players quit using spit tobacco. *Journal of the American Dental Association, 125,* 559-567.

6.3   EFFECTS OF A THIRD VARIABLE

At times you may be interested in summarizing more than two categorical variables. There are two different approaches for doing so; you may look at the marginal association between each pair of variables, or the conditional association of two variables at particular values of the third. We will illustrate both approaches for three dichotomous variables. Often, the conditional approach reveals interesting patterns of results and interactions of variables that are masked in the marginal analysis.

Marginal Association of Three Dichotomous Variables

The data for this example come from a study of crowding and antisocial behavior in 75 community areas in Chicago. Three characteristics of the communities are cross-classified to examine the relationships among socioeconomic status ("SES"), population density ("pop_dens"), and delinquency rate ("delinq"). Each variable is dichotomous, and has been coded as 1 = low and 2 = high. This data file is named "delinq.sav." The cross-tabulation of these three variables is shown in Table 6.1.

The marginal approach involves examining the association of each pair of variables. With three variables, there are three two-way combinations possible: SES _ pop_dens, SES _ delinq, and pop_dens _ delinq. The three tables appear in Figure 6.5. In order to have SPSS for Windows produce these tables, you need to create three separate 2 _ 2 tables using the steps in Section 6.1.

The first table shows the relationship between socioeconomic status and population density. The second and third tables portray the relationship of socioeconomic status with delinquency and population density with delinquency, respectively. Interpreting the first table, we see that 87.5% of the low SES communities have high population density, whereas only 17.1% of the high SES communities have high population density. Thus, low SES tends to be associated with high population density, and vice versa.

Table 6.1 Crosstabulation of 75 Communities by Delinquency, Population Density, and Socioeconomic Status

	Low SES		High SES	
	Low population density	**High population density**	**Low population density**	**High population density**
Low delinquency	3	2	27	3
High delinquency	2	33	2	3

ses * population density Crosstabulation

			population density		
			low	high	Total
ses	low	Count	5	35	40
		% within ses	12.5%	87.5%	100.0%
		% within population density	14.7%	85.4%	53.3%
	high	Count	29	6	35
		% within ses	82.9%	17.1%	100.0%
		% within population density	85.3%	14.6%	46.7%
Total		Count	34	41	75
		% within ses	45.3%	54.7%	100.0%
		% within population density	100.0%	100.0%	100.0%

SES * rate of juvenile delinquency Crosstabulation

			rate of juvenile delinquency		
			low	high	Total
SES	low	Count	5	35	40
		% within SES	12.5%	87.5%	100.0%
		% within rate of juvenile delinquency	14.3%	87.5%	53.3%
		% of Total	6.7%	46.7%	53.3%
	high	Count	30	5	35
		% within SES	85.7%	14.3%	100.0%
		% within rate of juvenile delinquency	85.7%	12.5%	46.7%
		% of Total	40.0%	6.7%	46.7%
Total		Count	35	40	75
		% within SES	46.7%	53.3%	100.0%
		% within rate of juvenile delinquency	100.0%	100.0%	100.0%
		% of Total	46.7%	53.3%	100.0%

Figure 6.5 Separate Two-Way Cross-tabulations for Three Dichotomous Variables

The second table indicates that low SES is associated with high rate of juvenile delinquency, and high SES with low rate of juvenile delinquency. The third table indicates that low population density is associated with low juvenile delinquency rate (88.2% of low population density communities have low delinquency rate). And, high population density is associated with high juvenile delinquency rate (87.8%).

population density * rate of juvenile delinquency Crosstabulation

			rate of juvenile delinquency		Total
			low	high	
population density	low	Count	30	4	34
		% within population density	88.2%	11.8%	100.0%
		% within rate of juvenile delinquency	85.7%	10.0%	45.3%
		% of Total	40.0%	5.3%	45.3%
	high	Count	5	36	41
		% within population density	12.2%	87.8%	100.0%
		% within rate of juvenile delinquency	14.3%	90.0%	54.7%
		% of Total	6.7%	48.0%	54.7%
Total		Count	35	40	75
		% within population density	46.7%	53.3%	100.0%
		% within rate of juvenile delinquency	100.0%	100.0%	100.0%
		% of Total	46.7%	53.3%	100.0%

Figure 6.5 *Continued*

Conditional Association of Three Dichotomous Variables

Examining three variables with the marginal approach is useful, but may hide some valuable information regarding how all three variables are related simultaneously. The conditional approach examines the association of two variables at each specific value ("layer") of the third. The relationship between two variables may be maintained, increased, decreased, or even reversed when a third variable is taken into account.

As an example, let's again look at the data on juvenile delinquency rates, SES, and population density. Table 6.2 displays the cell counts from the third table of Figure 6.5.

As we discussed, most of the low-delinquency communities are located in low-density areas, while most of the high-delinquency communities are in high-density areas. Now let us examine the relationship between delinquency and population when the third variable, socioeconomic status, is taken into account. To create a three-way cross-tabulation summary of the data, you need to use the "layer" option in the crosstabs dialog box (Fig. 6.1). Follow the same procedure detailed in Section 6.1, using "pop_dens" as the row variable, "delinq" as the column variable, and "SES" as the Layer 1 of 1 variable.

Your SPSS cross-tabular table should look like that in Figure 6.6.

Table 6.2 Cross-tabulation of Population Density by Delinquency

		DELINQ		
		Low	**High**	**All**
POP_DENS	Low	30	4	34
	High	5	36	41
	All	35	40	

population density * rate of juvenile delinquency * SES Crosstabulation

SES					rate of juvenile delinquency		Total
					low	high	
low	population density	low	Count		3	2	5
			% within population density		60.0%	40.0%	100.0%
			% within rate of juvenile delinquency		60.0%	5.7%	12.5%
			% of Total		7.5%	5.0%	12.5%
		high	Count		2	33	35
			% within population density		5.7%	94.3%	100.0%
			% within rate of juvenile delinquency		40.0%	94.3%	87.5%
			% of Total		5.0%	82.5%	87.5%
	Total		Count		5	35	40
			% within population density		12.5%	87.5%	100.0%
			% within rate of juvenile delinquency		100.0%	100.0%	100.0%
			% of Total		12.5%	87.5%	100.0%
high	population density	low	Count		27	2	29
			% within population density		93.1%	6.9%	100.0%
			% within rate of juvenile delinquency		90.0%	40.0%	82.9%
			% of Total		77.1%	5.7%	82.9%
		high	Count		3	3	6
			% within population density		50.0%	50.0%	100.0%
			% within rate of juvenile delinquency		10.0%	60.0%	17.1%
			% of Total		8.6%	8.6%	17.1%
	Total		Count		30	5	35
			% within population density		85.7%	14.3%	100.0%
			% within rate of juvenile delinquency		100.0%	100.0%	100.0%
			% of Total		85.7%	14.3%	100.0%

Figure 6.6 Cross-tabulation of 75 Communities by Delinquency, Population Density, and Socioeconomic Status

Figure 6.6 reveals a pattern that was not evident with the marginal associations shown in Figure 6.5 and Table 6.2. Among low SES areas, 60% of the low population density areas have low juvenile delinquency rate, and 94.3% of the high population density areas have a high delinquency rate. The story is not the same in high-SES areas, however. For instance, in high-SES areas that have high population density, the juvenile delinquency rate is split equally (50%) between low and high. Therefore, the pattern between population density and juvenile delinquency is different in different SES areas.

Chapter Exercises

6.1 Using the "titanic.sav" file, examine the relationship between "class" and "survival." Create a cross-tabulation table, with cell counts, and row, column, and total percentages.

    a. What percentage of first class passengers survived? Of the third class passengers?

    b. How many crew members were there? What percentage of the crew survived?

    c. Do you see a pattern regarding the relationship between survival and class?

6.2 Using the "fire.sav" data, use SPSS to do a cross-tabulation of race and sex and answer the following questions:

    a. What percentage of all firefighter applicants were minority females?

    b. Of the male applicants, were there more minority or white applicants?

    c. Calculate the phi-coefficient. How strong is the relationship between race and sex?

6.3 Using the "popular.sav" data, create a two-way frequency table of goals and gender

    a. What is the relationship, if any, between the two variables?

    b. Create a new crosstabulation table controlling for urbanicity. Does the location of the school affect the relationship between goals and gender? If so, how?

Part III

Probability

Chapter 7

Basic Ideas of Probability

The notion of chance is commonly used when defining probability. In statistical analysis, a random sample of individuals from the population is chosen in such a way that all possible sample sets have the same chance, or same probability, of being chosen. Although SPSS for Windows is designed primarily to be used for data analysis and not for evaluating probability functions per se, it is possible to demonstrate certain probability concepts with the program. This chapter illustrates "tossing a coin" and "rolling a die" using SPSS. All of the procedures discussed in this chapter involve *sampling with replacement;* using this approach, the probability of a particular outcome is not changed by of the outcome(s) that precede it.

7.1 PROBABILITY IN TERMS OF EQUALLY LIKELY CASES

This section demonstrates how to simulate equally likely outcomes of simple operations such as tossing a coin or dealing a card from a shuffled deck. The Bernoulli distribution is the distribution of a variable that can take one of two values — 0 or 1. There are several Bernoulli distributions, differing with respect to the probability associated with the values. If the probability is .5 that a random draw from this distribution will be a 1 (and .5 that it will be a 0), we can say that sampling from this distribution is the same as tossing a fair coin. For our purposes, obtaining a value of 1 corresponds to tossing a "head," and a value of 0 corresponds to a "tail."

To simulate coin tossing on SPSS:

1. Click on **File** from the menu bar.
2. Click on **New** from the pull-down menu.
3. Click on **Data** from the pull-down menu.
4. You need to compute a variable that has a value of 1 or 0 (a Bernoulli variable). SPSS will not permit you to compute any variable without having an active data set, however. To create such a data set, you must type some number (e.g., 999) in the first cell of the first column of the data file.
5. Click on **Transform** from the menu bar.
6. Click on **Compute** from the pull-down menu to open the Compute Variable dialog box (see Fig. 7.1).
7. Type in the name of the new variable (e.g., "coin") in the Target Variable box.
8. Click on **Random Numbers** in the Function group box and then double-click on **Rv.Bernoulli** in the Functions and Special Variables box. Notice that the function now appears in the Numeric Expression box.
9. Select a specific Bernoulli distribution by indicating the probability of obtaining a value of 1. Type **.5** where the question mark is.
10. Click on **OK** to run the procedure.

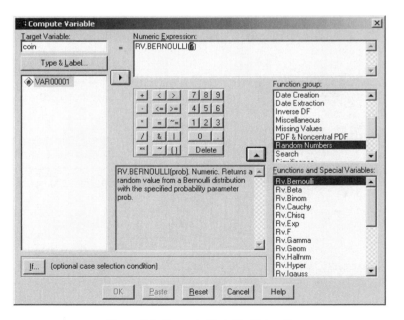

Figure 7.1 Compute Variable Dialog Box

Note that the value of the new variable, "coin," has a value of 1 (head) or 0 (tail). This represents the result of the coin toss.

You can also simulate rolling a die with a similar procedure. The random variable that corresponds to this is from a discrete uniform distribution, with values ranging from 1 to 6. SPSS has only the continuous uniform distribution as a random variable function. To make SPSS roll a die, follow steps 1–6 above, and then:

1. Type in the name of the new variable (e.g., "die") in the Target Variable box.

2. Click on **Arithmetic** in the Function group box and double-click on **Rnd** in the Functions and Special Variables box. This is the round function, which rounds continuous decimal numbers to integers.

3. Notice that "RND(?)" appears in the Numeric Expression box. Indicate the value or expression you wish to round. Click on Random Numbers in the Function group box and then double-click on **Rv.Uniform** in the Functions and Special Variables box. It should automatically replace the question mark from the "RND" expression.

4. Replace the question marks in the expression with **"1, 6"**.

5. Click on **OK** to run this procedure.

In the column labeled die in the data file, there will be an integer between 1 and 6, representing the roll of the die.

7.2   RANDOM SAMPLING; RANDOM NUMBERS

Drawing samples at random is often subject to biases (e.g., the deck was not shuffled adequately or the coin was slightly imperfect). By using numerical methods, the computer can simulate events with specific probabilities that are free from bias.

To generate several random numbers from a distribution, you can repeat the procedure in Section 7.1 several times. To do so, simply modify step (4) in Section 7.1 by entering a number for multiple cases for the variable. For instance, if you want to perform 10 coin tosses, you need to enter some number, all 999's for instance, in the first 10 cells of the first column. Then, compute the "coin" variable exactly as described above. You will obtain 10 values, each either 0 (tails) or 1 (heads), representing the result of each of ten coin tosses. Similarly, you may simulate 10 rolls of a die by using 1 and 6 as the minimum and maximum values, respectively, in the "RV.UNIFORM(min,max)" expression.

Chapter Exercises

7.1 Following the instructions in Section 7.2, simulate 20 coin tosses with SPSS.

    a. Did you obtain 10 heads (1's) and 10 tails (0's)? Why or why not?

    b. Would you have come closer to an equal split with 50 tosses? With 10 tosses? Why or why not?

7.2 Think of a number between 1 and 5. Direct SPSS to pick a number at random within the range 1 through 5. (Hint: use the uniform distribution).

    a. Did the two numbers coincide? What is the probability that they would be the same?

    b. Repeat this procedure 10 times. How many matches were there? How many did you expect to obtain (based on the law of probability)?

Chapter 8

Probability Distributions

This chapter demonstrates how to use SPSS for Windows to generate probability distributions. Probabilities of events can be described for categorical variables as shown in Chapter 7 (e.g., the likelihood that a coin toss will be a head), or numerically as described in this chapter (e.g., the likelihood that a person regularly eats 50 candy bars per year). We will focus on properties of the standard normal distribution, but SPSS can perform the same tasks for other probability distributions as well. Using SPSS, we demonstrate how to:

- find the probability associated with a given value of a standard normal variable;
- find the value of a standard normal variable associated with a given probability.

8.1 FAMILY OF STANDARD NORMAL DISTRIBUTIONS

Many numerical variables have distributions that look similar — a smooth curve in the shape of a bell. This shape is based on a probability distribution known as the normal distribution. The Central Limit Theorem tells us that whenever a variable is made up of many separate "components," each of which can have two or more values, the resulting variable will have approximately a normal probability distribution. The greater the number of components, the more perfectly normal the probability distribution for the variable will be. A normally shaped distribution with mean 0 and standard deviation 1 is called the standard normal distribution (see Chapter 4).

Finding Probability for a Given z-Value

When examining probability distributions, it is important to be able to determine the percentage of the distribution that lies within a certain interval. We know, for instance, that 50% of the area under the standard normal curve lies below the point 0. In other words, the probability that a random variable drawn from a standard normal distribution will be less than 0 is .50.

The cumulative distribution function in SPSS for Windows will compute these probabilities. The procedure for obtaining these computations is detailed below:

1. Click on **File** from the menu bar.
2. Click on **New** from the pull-down menu.
3. Click on **Data** from the pull-down menu.
4. SPSS will not permit you to compute a new variable without having an active data set. So, "activate" the Data window by typing some number (e.g., 999) in the first cell of the first column of the data file.
5. Click on **Transform** from the menu bar.
6. Click on **Compute** from the pull-down menu.
7. Type in the name of the new variable (e.g., "probilty") in the Target Variable box.
8. Click on **CDF & Noncentral CDF** in the Function group box and double-click on **Cdf.Normal** in the Functions and Special Variables box.
9. Notice the function in the Numeric Expression box. The first question mark (the "q" parameter) represents the point on the distribution for which you wish to obtain a cumulative probability estimate. For the first example, we will find the cumulative probability distribution for 0. Therefore, modify the expression to read "**CDF.NORMAL(0,0,1).**"
10. Click on **OK** to run the procedure.

SPSS will return the value of .5 as the value of the "probilty" variable.

Repeat this procedure, this time for the value 1. You can do so by starting at step (5) and changing the "q value" from 0 to 1. (Note: When you click on **OK** SPSS will prompt you to indicate whether or not you wish to "Change Existing Variable?") Then click on **OK** to run the procedure. The value of the "probilty" variable should change to .84, indicating that 84% of the distribution is below 1 standard deviation above the mean.

Finding a z-Value for a Given Probability

There may also be instances when you wish to determine the point on a probability distribution associated with a given cumulative probability. For instance, the z-value that has 50% of the standard normal distribution below it is 0, and a z-value that has 95% of the standard normal distribution below it is 1.64. The function in SPSS that will return the z-values for given probabilities is the inverse distribution function.

The procedure for obtaining this result is similar to that outlined previously. Follow steps 1–6 above, and then:

1. Type in the name of the new variable (e.g., "value") in the Target Variable box.
2. Click on **Inverse DF** in the Function group box and double-click on **Idf.Normal** in the Functions and Special Variables box.
3. To find the value on the standard normal distribution that has 95% of the values below it, modify the function in the Numeric Expression box by entering the value **.95** for the probability parameter, the value of **0** for the mean parameter, and the value of **1** for the stddev parameter.
4. Click on **OK** to run the procedure.

You should obtain the value of 1.64.

Chapter Exercises

8.1 Using SPSS, find the value on the standard normal distribution that has 2.3% of the distribution below it.

8.2 Sketch the curve of a normal distribution with mean 2 and standard deviation 1. How much of this distribution is below the value 2? Verify your answer using SPSS.

8.3 Estimate the value on the standard normal distribution that has 60% of the distribution below it. Use SPSS to evaluate your estimate.

8.4 Repeat Exercise 8.2 using a normal distribution with mean 0 and standard deviation 2.

8.5 Using SPSS, determine the proportion of the standard normal curve that is:

   a. above !1.

b. between !–2 and +2.

Hint: SPSS will not compute these areas directly; you must perform some simple computations by hand.

Chapter 9

Sampling Distributions

Statistical inference is a process of drawing conclusions about a population value (a parameter) based on a value computed from a random sample (a statistic). Each sample drawn from the population may yield a different statistic, however. The *sampling distribution* is the distribution of all possible values of a particular statistic, each with an associated probability. In this chapter, we use SPSS to simulate drawing random samples from a population, computing a particular statistic, and constructing its sampling distribution. In particular, we shall simulate the sampling distribution of a single observation drawn from a standard normal population distribution, the distribution of the sum of two observations drawn at random, and the distribution of the mean of 100 observations drawn from a population.

9.1 *SAMPLING FROM A POPULATION*

A random sample of a given size is a group of members of the population selected so that all groups of that size are equally likely to appear in the sample. Randomization ensures an equal chance of selection for every possible subset. A variable assessed for each member of the sample (selected through a random process) is called a random variable.

Random Samples

When we make SPSS toss a coin 10 times (Section 7.2), we are actually taking a random sample from a Bernoulli (two-valued) probability distribution. Now we shall repeat this procedure, but will draw a random sample of 50 observations from a standard normal distribution. Because the procedure is so similar to those described in Chapters 7 and 8, the following instructions are abbreviated.

1. Open a new Data window (see Section 7.1).
2. Enter "999" for the first 50 cases of the first column.
3. Click on **Transform** from the menu bar.
4. Click on **Compute** from the pull-down menu.
5. In the Compute dialog box, name the Target Variable "sample."
6. Highlight "Random Numbers" from Function Group box and "RV.Normal" from the Functions and Special Variables box and move it into the Numeric Expression box by clicking on the **up arrow button**.
7. Choose a standard normal distribution by replacing the first and second question marks with **0** and **1**, respectively.
8. Click on **OK**.

Now create a histogram of the "sample" variable (see Section 2.2 for details on creating histograms). It should resemble a standard normal distribution, but will differ somewhat because it is a sample. Figure 9.1 shows one possible histogram. This histogram would appear more normal if we had taken a larger number of samples. Because your random sample is different, your histogram will vary somewhat. Note, for instance, that the mean for this sample is −.11, which is close to the mean of 0, and the standard deviation of this random sample (0.90) is close to that of a standard normal distribution, 1.

9.2   SAMPLING DISTRIBUTION OF A SUM AND OF A MEAN

It is possible to use SPSS to simulate the sampling distribution of any statistic computed from a random sample from the population. In this section, we obtain the sampling distribution of the sum of two variables, each having a discrete uniform distribution with minimum of 1 and maximum of 6. This is analogous to constructing the sampling distribution resulting from rolling two dice repeatedly and tabulating the sum of the pips on each roll.

We will direct SPSS to roll dice, one at a time, for a total of 50 pairs of rolls. Next, we will compute the sum of each roll (e.g., the first roll for die one + the first roll for die two, etc.) and then examine the frequency distribution of this new variable.

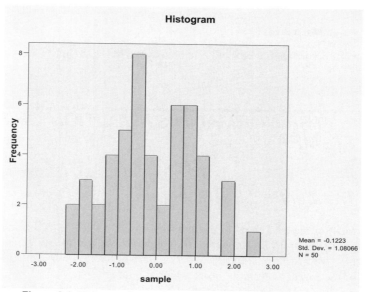

Figure 9.1 Histogram from a Standard Normal Distribution

1. Open a new Data window, and type "999" in the first 50 rows of the first column.
2. Click on **Transform** from the menu bar.
3. Click on **Compute** from the pull-down menu.
4. Name the target variable "die1."
5. In the Numeric Expression box, create the expression "**RND(RV.UNIFORM(1,6))**." (See Section 7.1.)
6. Click on **OK**.
7. Repeat steps 2–6 to compute another variable named "die2."
8. Click on **OK.**
9. Compute a "total" variable, which is the sum of the two sample variables. Click on **Transform** from the menu bar.
10. Click on **Compute** from the pull-down menu.
11. Click on the **Reset button** in the Compute Variable dialog box.
12. Next, name the target variable "**total**."
13. Click on the "die1" variable in the lower left box and move it to the Numeric Expression box with the **right arrow button**, click on the + from the calculator pad, then click on and move the "die2" variable to the Numeric Expression box with the **right arrow button**.
14. Click on **OK**.

15. Obtain a frequency distribution and histogram of the "total" variable (see Section 2.2).

Your output should be similar to that shown in Figure 9.2.

9.3   THE NORMAL DISTRIBUTION OF SAMPLE MEANS

The Central Limit Theorem

Most of the inferential procedures discussed in Part IV of this manual are based on the Central Limit Theorem (CLT) that states that the sampling distribution of the sample mean is approximately a normal distribution. This is true regardless of the parent distribution from which the samples are drawn, as long as the sample size (n) is large. Using the CLT allows us to make probability statements about a sample mean without actually observing the entire population from which it was drawn.

It is possible to use SPSS to illustrate the Central Limit Theorem. The process is not straightforward, but because the CLT is one of the most important principles of inferential statistics, working through this example may help you to understand the concepts more fully.

To illustrate this theorem, we will first have to obtain a random sample of size n (e.g., 50) from a specific distribution (e.g., discrete uniform(1,10)), and calculate the mean of the 50 observations. Then we will repeat this process many times (e.g., 99 more times), and inspect the frequency distribution and histogram of the sample means.

The procedures used to obtain a random sample from a specific distribution are given in Chapter 7. The tedious part of this process involves repeating the sampling 100 times. We have completed this step for you, and the results are saved in the data file "clt.sav." There are 100 variables, u1 to u100, which represent the 100 times that SPSS drew random samples of size 50. Therefore, at present we have a [50 rows × 100 columns] matrix representing [sample size × number of samples].

In order to get a histogram of the means, we first need to calculate the mean of each of the 100 samples (of the columns). We could direct SPSS to compute the mean of each of the "u" variables separately, but we would then have to manually input each of these means into another column. A less time consuming method is to transform the matrix in such a way as to make SPSS keep track of the means. To do so, we have to transpose the matrix — interchange the rows and the columns — and then compute the means. Open the "clt.sav" data file, and then:

total

		Frequency	Percent	Valid Percent	Cumulative Percent
Valid	2.00	2	4.0	4.0	4.0
	3.00	3	6.0	6.0	10.0
	4.00	5	10.0	10.0	20.0
	5.00	4	8.0	8.0	28.0
	6.00	8	16.0	16.0	44.0
	7.00	11	22.0	22.0	66.0
	8.00	10	20.0	20.0	86.0
	9.00	4	8.0	8.0	94.0
	11.00	3	6.0	6.0	100.0
	Total	50	100.0	100.0	

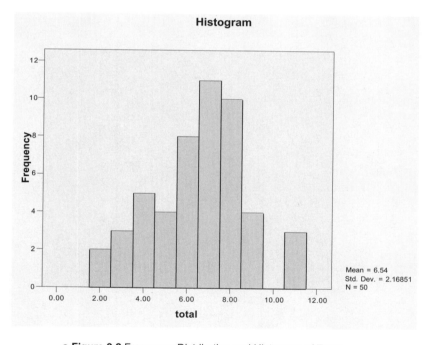

o **Figure 9.2** Frequency Distribution and Histogram of Total

1. Click on **Data** from the menu bar.
2. Click on **Transpose** from the pull-down menu.

3. Highlight all of the variable names (click on **u1**, hold the mouse button down, and drag down to the name of the last variable in the list).

4. Move the variable names to the Variable(s) box by clicking on the **upper right arrow button**.

5. Click on **OK**.

You should now have a transposed data file with a [100 × 50] matrix (excluding the first column). The rows now represent the 100 samples drawn, and the columns represent the 50 draws in each sample.

We can now compute the mean of each of the rows. To do this:

1. Click on **Transform** from the main menu bar.

2. Click on **Compute** from the pull-down menu.

3. Enter "mean" in the Target Variable box.

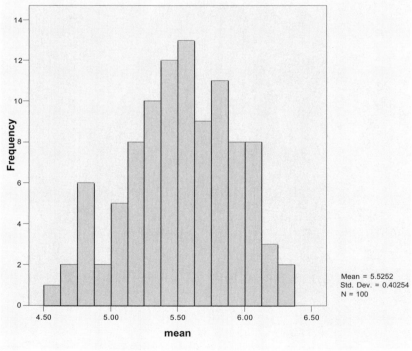

Figure 9.3 Histogram of Means

4. In the Numeric Expression box, create the expression: "**mean(var001 to var050)**." (The easiest way to do this is to type it in the box.)
5. Click on **OK**.

This should create a new variable, "mean," which contains the means of the 100 samples. Create a histogram of this new variable. Figure 9.3 displays this graph. Although the distribution of the means is not exactly normal, it is close to normal. Normality would be improved if the parent distribution were more normal-like or if we had drawn more than 100 samples.

Chapter Exercises

9.1 Use SPSS to:

   a. Obtain 10 random samples of size 8 from a standard normal distribution, and then compute the mean of each of these samples. Are all or any of the sample means equal to 0, the mean of the population? Sketch a histogram of these means and summarize the distribution.
   b. Repeat part (a), using 10 random samples of size 40.
   c. Compare your results in part (a) and part (b). What principle do these results illustrate?

9.2 Use SPSS to:

   a. Obtain three random samples of size 10 from a Bernoulli distribution with $p = .5$.
   b. Compute a new variable, which represents the sum of these three original variables.
   c. Based on probabilities, sketch the histogram you would expect to obtain for this composite variable.
   d. Create the histogram with SPSS and compare the actual results to those in part (c). How would you explain the differences?

Part IV

Inferential Statistics

Chapter 10

Answering Questions About Population Characteristics

Statistical inference is the process of using the characteristics of a sample to make statements about the population from which it is drawn. There are many forms of statistical inference, including interval estimation and hypothesis testing. These procedures can be used to:

- compute the range of plausible values for scores on an IQ test;
- determine whether mammals get, on average, 9 hours of sleep per day;
- test that the proportion of minority firefighter applicants is 26%, the same as the general population;
- determine whether reading scores for students exposed to a particular instruction program change over time.

10.1 AN INTERVAL OF PLAUSIBLE VALUES FOR A MEAN

The sample mean is a point estimate of the population mean, but it is usually not exactly equal to the population mean. The degree of accuracy can be indicated by reporting the estimate of the population mean as a range of values, that is, a confidence interval. The standard error, a measure of precision of the point estimate, is incorporated into the confidence interval.

In instances in which the population standard deviation (σ) is known, it can be used directly to compute the standard error and obtain the confidence interval. In most cases, however, the value of σ is not known and the sample standard deviation (s) must be substituted to give the estimate of the standard error of the mean, $s_{\bar{x}}$. SPSS for Windows bases its computations on the sample standard deviation, and we will illustrate only this situation.

We will compute the confidence interval for the mean number of calories per hot dog, using the data file "hotdog.sav."

1. Click on **Analyze** from the menu bar.

2. Click on **Descriptive Statistics** from the pull-down menu.

3. Click on **Explore**.

4. Click on the "calories" variable and move it into the Dependent List box with the **top right arrow button**.

5. Click on **Statistics** to open the Explore: Statistics dialog box.

6. Note that **Descriptives: Confidence Interval for Mean** is checked and 95% is the default. You may change this if you require a different level of confidence by moving your cursor to this box and typing in the desired level. For this exercise, we will maintain the default 95% level of confidence.

7. Click on **Continue**.

8. Click on **Statistics** in the Display box. This is an optional step, but recommended here because it suppresses tables that are unnecessary to address the question at hand.

9. Click on **OK**.

Your output will appear as shown in Figure 10.1. Note that the sample mean is 145.44 calories. The confidence interval is labeled "95% Confidence Interval for Mean" with a lower bound of 137.42 and an upper bound of 153.46. This indicates that we are 95% confident that the population mean is in the range 137.42 to 153.46 calories.

10.2   TESTING A HYPOTHESIS ABOUT A MEAN

The procedures for testing hypotheses about a mean differ slightly depending upon whether the standard deviation of the population (σ) is known or unknown. In this book, we illustrate the more common case, when the population standard deviation is unknown.

Case Processing Summary

| | Cases | | | | | |
| | Valid | | Missing | | Total | |
	N	Percent	N	Percent	N	Percent
CALORIES	54	100.0%	0	.0%	54	100.0%

Descriptives

			Statistic	Std. Error
CALORIES	Mean		145.4444	3.99857
	95% Confidence Interval for Mean	Lower Bound	137.4243	
		Upper Bound	153.4646	
	5% Trimmed Mean		146.0123	
	Median		145.0000	
	Variance		863.384	
	Std. Deviation		29.38339	
	Minimum		86.00	
	Maximum		195.00	
	Range		109.00	
	Interquartile Range		41.7500	
	Skewness		-.167	.325
	Kurtosis		-.691	.639

Figure 10.1 Explore Output with 95% Confidence Interval

The first step in hypothesis testing is generating two competing hypotheses for the question you are asking — the null hypothesis, denoted H_0, and the alternative hypothesis, denoted H_1. These hypotheses are about the value of the population mean, and are determined by the specific research question being asked. Hypothesis testing involves drawing a random sample from the population, computing the sample mean, and converting it to a *test statistic* that indicates how far the sample mean is from the hypothesized population mean. The test statistic is compared to percentage points of the appropriate probability distribution to decide if H_0 is maintained or rejected.

Validity Conditions

It is good practice to check the data for normality prior to conducting the test of significance. A visual way to inspect for normality is to plot a histogram of the variable. There is an option that directs SPSS to impose a normal curve on the graph, which makes it easier to evaluate the normality of the data. We will illustrate this using the calories variable of the "hotdog.sav" data file. You can obtain this graph using the Frequencies procedure as follows:

1. Click on **Graphs** from the menu.
2. Click on **Histogram** from the pull-down menu.
3. Click on and move the "calories" variable to the Variable box of the Histogram dialog box by clicking on the **right arrow button**.
4. Click on the **Display normal curve** option.
5. Click on **OK**.

Your output should look like that shown in Figure 10.2.

Although the distribution is not precisely normal, it is not highly skewed either. The histogram fits fairly well under the normal curve superimposed on the graph. Because the test is fairly "robust" with respect to normality, we conclude that it is an appropriate method for hypothesis testing in this instance.

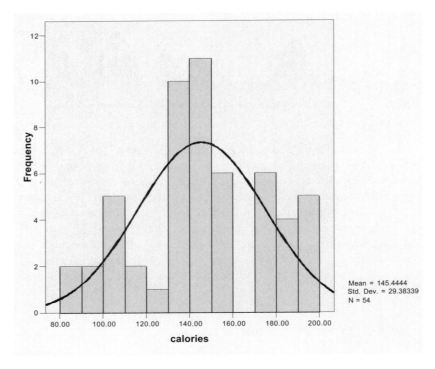

Mean = 145.4444
Std. Dev. = 29.38339
N = 54

Figure 10.2 Sample Histogram with Normal Curve Superimposed

Conducting the Hypothesis Test

When the standard deviation of the population is not known, you will need to estimate σ in order to compute the test statistic. SPSS computes the sample standard deviation, *s,* and uses it to calculate the appropriate test statistic (*t*). The procedure for using SPSS is the same regardless of whether you are making a one-tailed or a two-tailed test.

We will illustrate again using the calories variable from "hotdog.sav." Treating this sample as representative of the population of hotdogs, we will test the hypothesis that the population mean (represented as μ) number of calories is 150. The null and alternative hypotheses are H_0: μ = 150 and H_1: μ ≠ 150.

Before computing the test statistic, as the data analyst you must select your error level, α. This represents the probability of committing a Type I error; that is, the probability of rejecting the null hypothesis when it is true. In this example, we select an α level of .05. That is, we have chosen to accept a 5% chance of incorrectly rejecting the null hypothesis (incorrectly stating that hot dogs do not have, on average, 150 calories).

We can now use SPSS to determine the test statistic for this sample and use it to conduct the hypothesis test. Once the data file is open:

1. Click on **Analyze** from the menu bar.
2. Click on **Compare Means** from the pull-down menu.
3. Click on **One-Sample T Test** from the pull-down menu to open the One-Sample T Test dialog box (see Fig. 10.3).
4. Highlight the "calories" variable and move it to the Test Variable(s) box by clicking on the **right arrow button.**

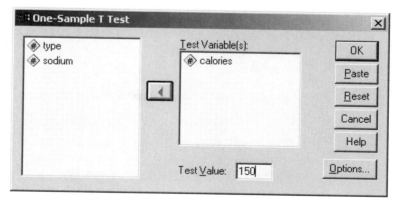

Figure 10.3 One-Sample T Test Dialog Box

One-Sample Statistics

	N	Mean	Std. Deviation	Std. Error Mean
CALORIES	54	145.4444	29.38339	3.99857

One-Sample Test

				Test Value = 150			
					Mean	95% Confidence Interval of the Difference	
	t	df	Sig. (2-tailed)	Mean Difference	Lower	Upper
CALORIES	-1.139	53	.260	-4.5556	-12.5757	3.4646

Figure 10.4. Sample Output for One-Sample T-Test

5.   In the Test Value box, enter the number 150.

6.   Click on **OK** to run the procedure.

The output should look like that in Figure 10.4.

From this listing we see that the mean of the 54 hotdog brands is 145.44 calories and the standard deviation is 29.38 calories (the same as was displayed in Figure 10.1). The t-statistic for the test is $t = -1.139$. This is found by computing the following:

$$t = \frac{\bar{x} - \mu_0}{\frac{s}{\sqrt{n}}} = \frac{145.44 - 150}{\frac{29.38}{\sqrt{54}}} = -1.139$$

Thus, the test statistic represents the number of standard errors the sample mean (here, 145.44 calories) is greater or less than the test mean (here, 150 calories).

In order to determine whether to reject the null hypothesis based on this test statistic, we must determine the probability of obtaining a value more extreme than our test statistic when the null hypothesis is true. This probability is called the P value. If the P value is less than our chosen α level, we reject the null hypothesis; if the P value is greater than our chosen α level, we do not reject the null hypothesis.

In Figure 10.4, the column labeled "Sig. (2-tailed)" is the P value for this test. Notice that the P value is listed as .260. So, if the null hypothesis were true (hot dogs have, on average, 150 calories), then the probability of obtaining a test statistic with an absolute value at least 1.139 is less than .260. This is greater than our chosen α level of .05, so we cannot reject the null hypothesis. We accept H_0 and conclude that the average number of calories in hot dogs does not differ from 150.

SPSS reports P values for two-tailed tests. If we were performing a one-tailed test, we would be concerned only with the upper (or lower) tail of the t-distribution. In order to obtain the correct achieved significance level, the P value produced by SPSS must be divided by 2. To reject the null hypothesis when conducting a one-tailed test, (a) $P/2$ must be less than α *and* (b) the sample mean must be in the direction specified by the alternative hypothesis (H_1).

Refer again to the hotdog example (Fig. 10.4). Suppose we want to test the null hypothesis H_0: $\mu \geq 150$ against the alternative H_1: $\mu < 150$, using $\alpha = .05$. The P value reported by SPSS is $P < .260$. We divide P by 2 and find that the achieved significance level is less than .130. This is not less than our α of .05. Thus, we do not reject H_0; we conclude that, on average, hot dogs do not have less than 150 calories.

Relationship Between Two-Tailed Tests and Confidence Intervals

There is a specific relationship between two-tailed tests and confidence intervals. If μ_0 lies within the $(1 - \alpha)$ confidence interval for the mean of a variable, the null hypothesis would not be rejected at the α level of significance. Conversely, if μ_0 is not within the interval, then the null hypothesis would be rejected. Return to Figure 10.1, and note that the 95% confidence interval contains the values of μ_0, 150 calories.

10.3   TESTING HYPOTHESES ABOUT A PROPORTION

The steps in testing hypotheses about a proportion are similar to testing hypotheses about a mean demonstrated in the previous section of this chapter. The null and alternative hypotheses are stated in terms of a probability, p, and a hypothesized value, p_0. The hypothesis is tested by computing the test statistic:

$$z = \frac{(\hat{p} - p_0)}{\sigma_{\hat{p}}}$$

where $\hat{p}$ is the sample proportion, and the standard error is

$$\sigma_{\hat{p}} = \sqrt{\frac{p_0(1 - p_0)}{n}}$$

We can test hypotheses about a proportion using the Binomial procedure in SPSS. Using the "fire.sav" data file, let us test the hypothesis that the proportion of white applicants is less than the national population percentage, which is approximately 74%. The hypotheses are H_0: $p \geq .74$ and H_1: $p < .74$. After you open the data file:

1. Click on **Analyze** from the menu bar.
2. Click on **Nonparametric Tests** from the pull-down menu.
3. Click on **Binomial** from the pull-down menu to open the Binomial Test dialog box (Fig. 10.5).
4. Click on and move the "race" variable to the Test Variable List box using the **right arrow button**.
5. Type .74 in the Test Proportion box. The .74 represents p_0.
6. Click on **OK** to run the procedure.

The output listing is displayed in Figure 10.6. The Test Proportion is the p_0 you entered in Step 5. The Obs. Prop. is the proportion of cases with value of 1 in the data file ($\frac{17}{28}$ in the example). (Note that SPSS always calculates the proportion using the value of the variable with the larger number of cases. Therefore, you must be careful to enter the appropriate test proportion.)

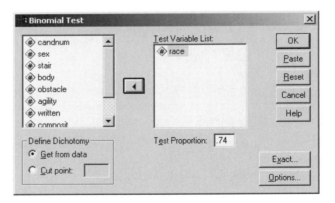

Figure 10.5 Binomial Test Dialog Box

Binomial Test

		Category	N	Observed Prop.	Test Prop.	Asymp. Sig. (1-tailed)
race	Group 1	white	17	.61	.74	.086[a,b]
	Group 2	minority	11	.39		
	Total		28	1.00		

a. Alternative hypothesis states that the proportion of cases in the first group < .74.

b. Based on Z Approximation.

Figure 10.6 Sample Output for Binomial Test

The printout also contains the P value using the normal approximation to the binomial distribution. SPSS employs a continuity correction for $\hat{p}$, adding $\dfrac{1}{2n}$ to the sample proportion. The "corrected" sample proportion is $\dfrac{17}{28} + \dfrac{1}{56} = .625$, and the test statistic is

$$z = \frac{(.625 - .74)}{\sqrt{\dfrac{.74 \times .26}{28}}} = -1.39$$

The corresponding P value from the approximation to the standard normal distribution is $P = .086$. Note that the footnotes indicate this is a one-tailed P value. If we were using an α level of .05, we would accept the null hypothesis that the proportion of white applicants is greater than or equal to .74.

10.4   PAIRED MEASUREMENTS

There are two main types of paired measurements that occur in statistics. One involves two measurements being made on one unit of observation at two times, such as body weight before and after a diet, or the scores made on a college admissions test before and after a preparatory class. Matched samples are also used to choose individuals with similar characteristics but assigned to different experimental conditions. We shall demonstrate how to use SPSS to test hypotheses about the mean of a population of differences (often referred to as difference scores), and hypotheses about the equality of proportions from paired measurements.

Testing Hypotheses About the Mean of a Population of Differences

SPSS conducts a test of whether the mean of a population of difference scores is equal to 0. We illustrate with the "reading.sav" data file. This file contains reading scores for 30 students obtained on the same test administered before and after second grade. We want to determine whether reading skill increases, on average, throughout second grade. Suppose we choose an α level of .05. After opening the data file:

1. Click on **Analyze** from the menu bar.
2. Click on **Compare Means** from the pull-down menu.
3. Click on **Paired-Samples T Test** from the pull-down menu. This opens the Paired-Samples T Test dialog box (see Fig. 10.7).
4. Click on the "before" variable. It will appear in the Current Selections box as Variable 1.
5. Click on the "after" variable. It will appear in the Current Selections box as Variable 2.
6. Move the paired variables into the Paired Variables box by clicking on the **right arrow button**.
7. Click on **OK** to run the procedure.

Figure 10.8 displays the output from this procedure. The top portion lists the means, standard deviations, and standard errors of reading scores before and after second grade. The scores increased, on average, from 1.52 to 2.03 points.

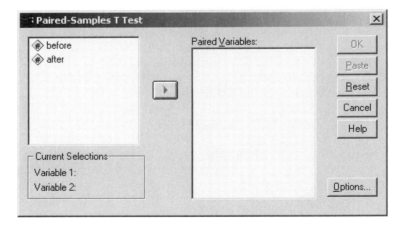

Figure 10.7 Paired-Samples T Test Dialog Box

The middle portion includes the sample correlation coefficient between the pre-test and post-test (.573) and a test of significance of the correlation ($P < .001$). This is not a component of the hypothesis test we are conducting in this example.

The lowest portion of the output contains information regarding the test of the hypothesis that the mean difference is equal to 0. The mean difference, $-.5100$, is equivalent to $1.5233 - 2.0333$. The table also displays the standard deviation and the standard error of the difference. The t statistic is

$$t = \frac{-.5100 - 0}{\frac{.492}{\sqrt{30}}} = -5.683 \ .$$

The two-tailed P value is less than .0005 (and is rounded to .000 in SPSS). However, we are conducting a one-tailed test because we began by speculating that reading scores would increase. Therefore, we must compare the $P/2$ value to α and verify that the sample post-test mean is higher than the sample pretest mean. In this case, $P/2 < .00025$, which is less than our selected significance level, .05. In addition, mean score after second grade is higher than that before second grade. Consequently, we reject the null hypothesis and conclude that reading skills increase during grade 2.

Paired Samples Statistics

		Mean	N	Std. Deviation	Std. Error Mean
Pair 1	Reading Score Before 2nd Grade	1.5233	30	.27628	.05044
	Reading Score After 2nd Grade	2.0333	30	.59442	.10853

Paired Samples Correlations

		N	Correlation	Sig.
Pair 1	Reading Score Before 2nd Grade & Reading Score After 2nd Grade	30	.573	.001

Paired Samples Test

		Paired Differences					t	df	Sig. (2-tailed)
		Mean	Std. Deviation	Std. Error Mean	95% Confidence Interval of the Difference Lower	Upper			
Pair 1	Reading Score Before 2nd Grade - Reading Score After 2nd Grade	-.5100	.49155	.08974	-.6935	-.3265	-5.683	29	.000

Figure 10.8 Sample Output for t Test for Paired Samples

Testing the Hypothesis of Equal Proportions

You may also examine changes in time for dichotomous variables by testing the equality of two proportions obtained from a single sample. Some textbooks refer to this a "turnover table," and SPSS labels it the McNemar test for correlated proportions. Using SPSS, the procedure produces a chi-square test statistic.

We will illustrate this using the "war.sav" file. This file contains the data from a study examining changes in attitudes regarding the likelihood of war. In both June and October of 1948, subjects were asked to indicate whether or not they expected a war in the next ten years. The "war.sav" data file is coded so that a 2 represents "Expects War" and a 1 represents "Does Not Expect War." Let us test this at the .05 level of significance.

To test the hypothesis of equal proportions, open the data file and:

1. Click on **Analyze** from the menu bar.
2. Click on **Nonparametric Tests** from the pull-down menu.
3. Click on **2 Related Samples** from the pull-down menu. This opens the Two-Related-Samples Tests dialog box (Fig. 10.9).
4. Click on the variable name "June." It will appear as Variable 1 in the Current Selections box.
5. Click on the variable name "October." It will appear as Variable 2 in the Current Selections box.
6. Move the paired variables to the Test Pair(s) List box by clicking on the **right arrow button**.
7. In the Test Type box, click off the **Wilcoxon** box and click on the **McNemar** box.
8. Click on **OK**.

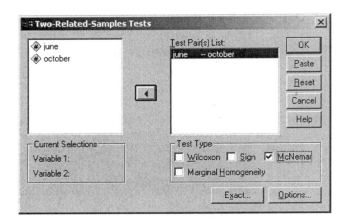

Figure 10.9 Two-Related-Samples Test Dialog Box

june & october

june	october 1	2
1	194	45
2	147	211

Test Statistics[b]

	june & october
N	597
Chi-Square[a]	53.130
Asymp. Sig.	.000

a. Continuity Corrected

b. McNemar Test

Figure 10.10 McNemar Test for Likelihood of War

The output (Fig. 10.10) displays the data in the two-way frequency table. We see that 45 people who did not think there would be war when questioned in June changed their minds for the October polling. Similarly, there were 147 individuals who in June thought there would be war, but in October did not believe this to be the case. The listing also reports the chi-square test statistic (53.130) and the corresponding P value ($P < .0005$). Therefore, we would reject the null hypothesis of equal proportions at any α level greater than .0005.

Note that the McNemar test is a two-tailed test. To perform a one-tailed test, compare $P/2$ to α and check that the sample proportions are in the direction indicated by the alternative hypothesis.

Chapter Exercises

10.1 Use SPSS and the "noise.sav" data file to:

    a. Test the null hypothesis that the average speed of automobiles is equal to 35 mph. Use $\alpha = .05$, and state the test statistic and conclusion.

    b. Would your decision change if you used $\alpha = .10$?

10.2 Use SPSS and the "football.sav" data file to complete the following:

   a.   Find the 90% confidence interval for the number of points by which football games are won. State the sample mean as well.

   b.   Would you reject the null hypothesis H_0: $\mu = 10$ points using $\alpha = .10$? (Hint: refer to your conclusions in part (a).)

   c.   What are your conclusions for the hypothesis H_0: $\mu \le 10$ points, using $\alpha = .10$?

   d.   What is the P value for the hypothesis in part (c)?

10.3   Use SPSS and the "bodytemp.sav" data file, which contains body temperature (in degrees Fahrenheit) for 130 adults, to complete the following:

   a.   What is the P value for the test of the null hypothesis that the average body temperature is 98.6 Fahrenheit?

   b.   Would you reject the hypothesis at $\alpha = .05$? $\alpha = .01$?

10.4   Use the "popular.sav" data file to complete the following:

   a.   Estimate the proportion of students who stated that their goal was to "make good grades."

   b.   Using $\alpha = .10$, test the hypothesis that the proportion of students wanting to make good grades is greater than 50%. (Hint: You my need to recode the "goals" variable, or use the cut point option.)

10.5   Using the "conform.sav" data file, answer the following questions:

   a.   Are wives more comformist, on average, than their husbands (use $\alpha = .05$)?

   b.   What is the minimum α for which you would reject the null hypothesis indicated in part (a)?

Chapter 11

Differences Between Two Populations

This chapter focuses on the comparison of two populations. This type of comparison is used to answer questions such as:

- On average, do females and males have the same body temperature?
- Do people who take one aspirin tablet every morning have lower risk of heart attack?
- Is one toothpaste generally superior to another in fighting cavities?
- On average, do graduate students indicate liking statistics more than undergraduate students?

To answer these questions, a sample is drawn from each population (e.g., males and females) and the means (or proportions or medians) of the samples are compared. If the difference between the samples is greater than expected as a result of sampling error, we conclude that the populations are distinct. As with the hypothesis testing in Chapter 10, procedures differ depending upon whether the population standard deviations are known. In this book, we consider the more likely instance, when the population standard deviations are unknown.

11.1 COMPARISON OF TWO INDEPENDENT MEANS

In comparing population means, we rarely know the population standard deviations of the two distributions. They may, however, be estimated to

123

determine the standard error of the difference of the means. Testing procedures are slightly different based on whether or not it can be assumed that the two standard deviations are equal. SPSS performs the test for both types of conditions. This section discusses the more common situation in which the standard deviations are treated as equal.

The "bodytemp.sav" data file contains information on body temperature and pulse rate for 130 male and female adults. Suppose we wish to determine whether males and females differ, on average, in normal body temperature. The null and alternative hypotheses are H_0: $\mu_{female} = \mu_{male}$ and H_1: $\mu_{female} \neq \mu_{male}$. We shall test the hypothesis at the 5% level.

We can now use SPSS to conduct the test as follows. After opening the data file:

1. Click on **Analyze** from the menu bar.
2. Click on **Compare Means** from the pull-down menu.
3. Click on **Independent Samples T-Test** from the pull-down menu to open the Independent-Samples T Test dialog box (Fig. 11.1).
4. Click on and move the "temp" variable the Test Variable(s) box using the **upper right arrow button**.
5. Click on and move the "sex" variable to the Grouping Variable box using the **lower right arrow button**.
6. Notice that two question marks appear in parentheses after the variable "sex." This signifies that you need to indicate the two values of the class variable for which you wish to calculate mean differences. To do so, click on **Define Groups** to open the Define Groups dialog box (see Fig. 11.2).
7. In our example, females are coded 0 and males are coded 1. Therefore, enter these numbers in the Group 1 and Group 2 boxes. (The cut point option is used if there are more than two values of the grouping variable.)
8. Click on **Continue** to close the dialog box.
9. Click on **OK** to run the procedure.

The output is displayed in Figure 11.3. The upper portion of the listing displays summary information (n's, means, standard deviations, and standard errors) for each of the samples. In this case, females had an average of 98.394°F and males had an average temperature of 98.105°F. The difference is 98.394 − 98.105 = .289°F. The figure is listed in the lower table in the "Mean Difference" column.

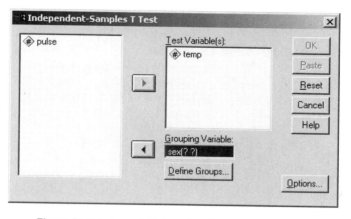

Figure 11.1 Independent-Samples T Test Dialog Box

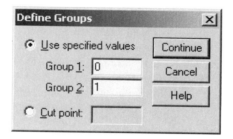

Figure 11.2 Define Groups Dialog Box

Group Statistics

	sex	N	Mean	Std. Deviation	Std. Error Mean
body temperature	female	65	98.394	.7435	.0922
(degrees Fahrenheit)	male	65	98.105	.6988	.0867

Independent Samples Test

		Levene's Test for Equality of Variances		t-test for Equality of Means							
									95% Confidence Interval of the Difference	
		F	Sig.	t	df	Sig. (2-tailed)	Mean Difference	Std. Error Difference	Lower	Upper
body temperature (degrees Fahrenheit)	Equal variances assumed	.061	.805	2.285	128	.024	.289	.1266	.0388	.5396
	Equal variances not assumed			2.285	127.510	.024	.289	.1266	.0388	.5396

Figure 11.3 T Test for Independent Samples

This table also displays two different t-statistics, one based on the assumption of equal variances, the other assuming unequal variances. We will only consider the equal variances case. The test statistic is $t = 2.285$.

The t-statistic is compared with significance points from the t-distribution with $130 - 2 = 128$ degrees of freedom. This is done by SPSS, resulting in the printed P value (in the Sig. (2-tailed) column). Since $P = .024$ is less than .05, the null hypothesis is rejected, and we conclude that, on average, females do not have a higher body temperature than do males.

The output also includes a 95% confidence interval for the mean difference. That is, the difference between average temperature for men and women is between 0.04 and 0.54 degrees with 95% confidence. Because 0 (representing no difference in average temperature between women and men) is not in the range, the results of the significance test are confirmed.

One-Tailed Tests

The SPSS procedure for conducting a one-tailed test is the same as that for a two-tailed test; it differs only in how the P value is used. Because the reported P value is for a two-tailed test, we must compare $P/2$ to α, and also verify that the sample means differ in the direction supported by the alternative hypothesis. In the example, $.012 < .05$, and the sample mean for males is less than the sample mean for females. Thus, we would reject H_0: $\mu_{female} \leq \mu_{male}$ in favor of H_1: $\mu_{female} > \mu_{male}$ at the 5% level of significance.

Chapter Exercises

11.1 Use the "enroll.sav" data file and SPSS to test whether there is a significant difference in the racial disproportion index between districts with high and low percentages of students paying full price for lunch, as follows:

    a. Recode the "pct_lnch" variable into a dichotomous variable, with 51% as the split point. (That is, values less than 51% will constitute the "low" group, and all other values the "high" group.)

    b. Would you perform a one- or two-tailed test? Why?

    c. Based on your response to part (b), state the null and alternative hypothesis.

    d. Use SPSS to conduct the test. State the value of the test statistic and the P value. (Assume equal variances.) Is H_0 rejected if $\alpha = .05$? If $\alpha = .01$?

11.2 Use the "football.sav" data file to explore the relationship between the type of team (home or away) winning games and number of points by which the game is won.

    a. State the null and alternative hypotheses for testing whether games that are won by the home team are won by, on average, more points than games that are won by the visiting team.

    b. Assuming equal variances, use SPSS to conduct the test. What are your conclusions using $\alpha = .01$? Give the P value for the test.

11.3 The "cars.sav" data file contains information on cars collected from a university parking lot. The information collected was: age, color and owner (faculty/staff or student).

    a. State the null and alternative hypotheses for testing whether students drive cars that are, on average, the same age as faculty/staff.

    b. Assuming equal variances, use SPSS to conduct the test. What are your conclusions using $\alpha = .05$? Give the P value for the test.

Chapter 12

Inference on Categorical Data

Significance tests of categorical variables involve the comparison of a set of observed frequencies with frequencies specified by a hypothetical distribution. We may ask, for instance:

- Do people have the same probability of dying in the month in which they were born as in any other month?
- Is there a relationship between race/ethnicity and political affiliation?
- Is choice of occupation related to one's sex?

This chapter details how to use SPSS to calculate goodness of fit with equal and unequal probabilities, to perform a chi-square test of independence, and to calculate measures of association between categorical variables, for example, the phi coefficient, coefficient lambda, and coefficient gamma.

12.1 TESTS OF GOODNESS OF FIT

When data consist of one categorical variable, it is often informative to ask whether the proportions of responses in the categories conform to a particular pattern. The procedure for addressing such questions is called a *goodness of fit test*.

Equal Probabilities

In this illustration, we will use the data in the "death.sav" data file to test the hypothesis that there are equal probabilities of death occurring in one's birth month or any other month of the year. The null hypothesis is H_0: $p_1 = p_2 = \ldots = p_{12} = 1/12$. Each entry in the data file is a number indicating the individual's month of death relative to the month of birth; for example, –6 indicates that the month of death is 6 months prior to the month of birth, 0 indicates that both months are the same, and so on. We will test this with an α level of .01.

By default, SPSS calculates the chi-square statistic to test the hypothesis of equal proportions. After you have opened the data file:

1. Click on **Analyze** from the menu bar.
2. Click on **Nonparametric tests** from the pull-down menu.
3. Click on **Chi-Square** to open the Chi-Square Test dialog box (see Fig. 12.1).
4. Click on the variable name ("month") and the **right arrow button** to move it into the Test Variable List box.
5. Click on **OK**.

The output should appear as shown in Figure 12.2.

Because we hypothesized that the chance of dying in any month is equal in proportion, we see that the expected number of individuals who died during each of the 12 months is 348/12 = 29. The test statistic is 22.07 with 11 degrees of freedom. The P value of .024 leads us to accept H_0 at the 1% level and conclude that people have an equally likely chance of dying in the month in which they were born as in any other month.

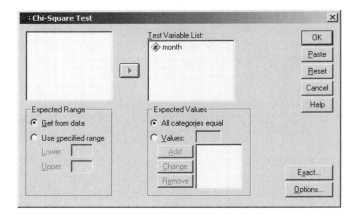

Figure 12.1 Chi-Square Test Dialog Box

MONTH

	Observed N	Expected N	Residual
-6	24	29.0	-5.0
-5	31	29.0	2.0
-4	20	29.0	-9.0
-3	23	29.0	-6.0
-2	34	29.0	5.0
-1	16	29.0	-13.0
0	26	29.0	-3.0
1	36	29.0	7.0
2	37	29.0	8.0
3	41	29.0	12.0
4	26	29.0	-3.0
5	34	29.0	5.0
Total	348		

Test Statistics

	MONTH
Chi-Square[a]	22.069
df	11
Asymp. Sig.	.024

a. 0 cells (.0%) have expected frequencies less than 5. The minimum expected cell frequency is 29.0.

Figure 12.2 Chi-Square Test of Goodness of Fit

Probabilities Not Equal

It is also possible to conduct goodness-of-fit tests when the proportions in the categories are hypothesized to be unequal. For example, car manufacturers and dealerships keep records on popularity of colors for automobiles. These records indicate that the color distribution is: blue 9.3%, sliver/gray 26.0%, red 13.4%, black 10.2%, green 10.4%, white 14.7%, brown/gold/other 16.0%. Suppose we want to test whether taste in colors on college campuses is the same as the national average. The "cars.sav" datafile contains the results of a survey of college parking lots.

The null hypothesis for this test is H_0: $p_{blue} = .093$, $p_{silver} = .260$, $p_{red} = .134$, $p_{black} = .102$, $p_{green} = .104$, $p_{white} = .147$, $p_{brown/other} = .160$. To test this hypothesis at the .05 level of significance, open the "cars.sav" data file and follow steps 1–4 above, clicking on the variable "color." Then:

1. Click on **Values** in the Expected Values box.

2. Enter the value (proportion) that you hypothesize for the first category of your variable. In this example, blue is coded "1" so enter .093 in the Value box and click on **Add**.

3. Enter the value for the next category of your variable. Silver is coded "2," so enter .260 and click on **Add**.

4. After entering all of the expected values, click on **OK**.

The output should appear as shown in Figure 12.3.

The Expected N column indicates the expected number of cars of each color if the hypothesized distribution were true. For example, 9.3% of the sample of 64 cars is 6.0. The actual number of blue cars in the sample was 7 (see the Observed N column). The test statistic is a measure of the magnitude of the differences between observed and expected N's. In this example, $\chi^2 = 2.060$, $P = .914$. Thus, using a 5% significance level, we do not reject the null hypothesis of independence and conclude that choice of car colors on college campuses does not depart from the national figures.

car color

	Observed N	Expected N	Residual
blue	7	6.0	1.0
grey	15	16.6	-1.6
red	7	8.6	-1.6
black	9	6.5	2.5
green	8	6.7	1.3
white	8	9.4	-1.4
brown	10	10.2	-.2
Total	64		

Test Statistics

	car color
Chi-Square[a]	2.060
df	6
Asymp. Sig.	.914

[a.] 0 cells (.0%) have expected frequencies less than 5. The minimum expected cell frequency is 6.0.

Figure 12.3 Goodness-of-Fit Test for Unequal Proportions

12.2   CHI-SQUARE TESTS OF INDEPENDENCE

When a study concerns the relationship of two categorical variables, SPSS can be used to test whether the variables are independent in the population. The pattern of observed frequencies in the sample is compared to the pattern that would be expected if the variables were independent. If the two patterns are quite different, we conclude that the variables are related in the population.

To illustrate, we shall use the data in the "popular.sav" data file that contains the results of a survey administered to 478 middles school children. One of the survey items directed the children to pick their most important goal: make good grades, be popular, or be good in sports. Consider the hypothesis that goal is independent of gender among children, using a 5% error rate.

To test this hypothesis, open the "popular.sav" data file and follow these steps:

1. Click on **Analyze** from the menu bar.
2. Click on **Descriptive Statistics** from the pull-down menu.
3. Click on **Crosstabs** to open the Crosstabs dialog box.
4. Click on the name of the row variable ("gender") and the **top right arrow button**.
5. Click on the name of the column variable ("goals") and the **middle right arrow button**.
6. Click on the **Cells** button to open the Crosstabs: Cell Display dialog box.
7. Click on **Row** in the Percentages box to indicate that you want percentages by gender (the row variable).
8. Click on **Continue**.
9. Click on the **Statistics** button to open the Crosstabs: Statistics dialog box (see Fig. 12.4).
10. Click on **Chi-Square**.
11. Click on **Continue**.
12. Click on **OK**.

Relevant output is displayed in Figure 12.5.

The test statistic we require is the one labeled Pearson under the Chi-Square heading. For these data the test statistic is $\chi^2 = 21.455$ with 2 degrees of freedom. The P value (Asymp. Sig. (2-sided) column) is less than .0005 (and is rounded to .000), leading us to conclude that choice of goal is not independent of gender. Looking at the percentages in the Cross-tabulation table, we see that girls are more likely than boys to choose being popular as a goal (in the sample, 36.3% compared to 22.0%), while boys have a larger tendency than girls to select being good in sports as a goal.

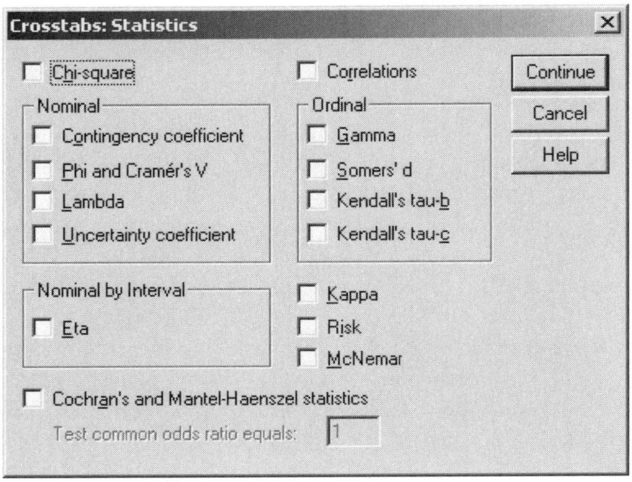

Figure 12.4 Crosstabs: Statistics Dialog Box

GENDER * GOALS Crosstabulation

			GOALS			
			make good grades	be popular	be good in sports	Total
GENDER	girl	Count	130	91	30	251
		% within GENDER	51.8%	36.3%	12.0%	100.0%
	boy	Count	117	50	60	227
		% within GENDER	51.5%	22.0%	26.4%	100.0%
Total		Count	247	141	90	478
		% within GENDER	51.7%	29.5%	18.8%	100.0%

Chi-Square Tests

	Value	df	Asymp. Sig. (2-sided)
Pearson Chi-Square	21.455[a]	2	.000
Likelihood Ratio	21.769	2	.000
Linear-by-Linear Association	4.322	1	.038
N of Valid Cases	478		

a. 0 cells (.0%) have expected count less than 5. The minimum expected count is 42.74.

Figure 12.5 Chi-Square Test of Independence Between Gender and Goals

There is additional information available and provided by the Crosstabs procedure. For instance, by default SPSS prints the minimum expected frequency (42.74) and the number (and percent) of cells that have an expected value less than 5. If many of the cells (e.g., 20% or more) have expected values below 5, the data analyst should consider combining some of the response categories. This is accomplished using the Recode procedure (Chapter 1) prior to conducting the chi-square test.

As an option, SPSS will print the expected values for each cell. To obtain the expected frequencies, follow the steps outlined above, but remembering to select the **Expected** option in the Crosstabs: Cell Display dialog box (refer to steps 6 and 7).

12.3 MEASURES OF ASSOCIATION

As with numerical variables, when two categorical variables are not independent, they are said to be correlated or associated with one another. It is possible to calculate an index that measures the degree of association between the two variables. The type of index that is most appropriate depends upon the nature of the variables (nominal or ordinal) and whether one of the variables can be considered a predictor of the other. SPSS labels a correlation in which there is no predictor variable as symmetrical, and a correlation in which there is a predictor as directional.

The Phi coefficient (ϕ) is appropriate for two nominal, dichotomous, variables and symmetrical situations. Nominal variables and a directional situation call for a lambda coefficient (λ). The gamma coefficient (γ) is a correlation for ordinal variables in which there is no prediction (symmetrical); Somer's d is used for ordinal variables in which it is appropriate to predict one variable from the other (directional).

The procedure for computing these coefficients with SPSS is virtually identical. We shall illustrate with "titanic.sav" by computing coefficient lambda for the relationship between class (first, second, third, crew) and survival (no, yes). We consider both these variables nominal, and predict survival from class. To use SPSS to compute this measure of association, open the data file and repeat the Crosstabs procedure described in Section 12.2, with the following features: use class as the row variable and survived as the column variable; select the **Lambda** coefficient instead of the Chi-square statistic in the Crosstabs: Statistics dialog box (see Fig. 12.4).

The output will appear as shown in Figure 12.6. The table shows that the coefficient for survived as the dependent variable is 0.114, which is weak. The test of significance for the λ-coefficients are displayed as t-statistics (Approx. T) and P values (Approx. Sig.). In this example, $P < .0005$, so we conclude that the variables are related.

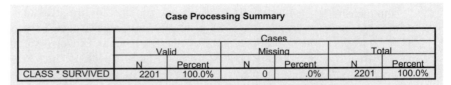

Case Processing Summary

	Cases					
	Valid		Missing		Total	
	N	Percent	N	Percent	N	Percent
CLASS * SURVIVED	2201	100.0%	0	.0%	2201	100.0%

CLASS * SURVIVED Crosstabulation

Count

		SURVIVED		Total
		no	yes	
CLASS	crew	673	212	885
	first	122	203	325
	second	167	118	285
	third	528	178	706
Total		1490	711	2201

Directional Measures

			Value	Asymp. Std. Error[a]	Approx. T[b]	Approx. Sig.
Nominal by Nominal	Lambda	Symmetric	.040	.009	4.514	.000
		CLASS Dependent	.000	.000	.[c]	.[c]
		SURVIVED Dependent	.114	.024	4.514	.000
	Goodman and Kruskal tau	CLASS Dependent	.025	.004		.000[d]
		SURVIVED Dependent	.087	.013		.000[d]

a. Not assuming the null hypothesis.

b. Using the asymptotic standard error assuming the null hypothesis.

c. Cannot be computed because the asymptotic standard error equals zero.

d. Based on chi-square approximation

Figure 12.6 Measure of Association Between Class and Survival Using Lambda

Chapter Exercises

12.1 Using the "titanic.sav" datafile:

    a. Perform a chi-square test to determine whether class and survival are independent. Use a significance level of .05; state the test statistic, P value, and your conclusion.

    b. If you rejected the null hypothesis in part a, describe the nature of the relationship between the two variables.

12.2 The "popular.sav" data file contains information on students' ratings of the importance of such things as making money, looks, sports, and grades in their lives. The students' responses are on a 4-point ordinal scale, where 1 = most important and 4 = least important.

a. Perform a chi-square test to determine whether gender and importance of money are independent. Use an α level of .05. What is your conclusion?

b. Perform a second test to determine whether gender and importance of looks are independent. Use an α level of .05. What is your conclusion?

12.3 Use the data on interventions aimed at reducing tobacco use among baseball players ("spit.sav") to answer the following:

a. What correlation coefficient is appropriate for determining the relationship of intervention and outcome? Why?

b. Compute the appropriate correlation coefficient. Are the variables related, based on a .05 level of error? If so, describe the nature of the relationship.

Chapter 13

Regression Analysis: Inference on Two or More Numerical Variables

It is often necessary to examine the directional relationship between two variables. For example:

- Does the amount of exposure to radioactive materials affect the cancer mortality rate?
- Do SAT scores predict college success?
- Does the speed on highways affect noise level?

Or, it may be necessary to examine the effect of two or more independent variables on one dependent variable. For example:

- Are the number of grams of carbohydrates and of fiber related to the number of calories in breakfast cereal?
- Do SAT scores and high school GPA predict college success?
- Are age, cholesterol level, and amount of exercise a person gets related to his or her chance of having a heart attack?

This chapter describes how to use SPSS for Windows to perform linear regression analysis to estimate statistical relationships from a sample of data. Because a scatter plot and correlation coefficient are indispensable in interpreting regression results, procedures for obtaining these are reviewed as well.

13.1 THE SCATTER PLOT AND CORRELATION COEFFICIENT

Two important steps in regression analysis involve examining a scatter plot of two variables and calculating the correlation coefficient. Although both of these procedures are described in Chapter 5, we will illustrate them here using the "cancer.sav" data file. In this example, we wish to examine the relationship between the amount of exposure to radioactive materials and cancer mortality rate. To create a scatter plot of these variables, open the data file and:

1. Click on **Graphs** from the menu bar.
2. Click on **Scatter/Dot** to open the Scatterplot dialog box.
3. Click on **Simple Scatter** and then on **Define** to open the Simple Scatterplot dialog box.
4. Click on the name of the independent variable (x) that you wish to examine ("expose") and move it to the X Axis box using the **second right arrow button**.
5. Click on the name of the dependent variable (y) that you wish to examine ("mortalit") and move it to the Y Axis box using the **top right arrow button**.
6. Click on **OK**.

The correlation coefficient can be calculated by using the following commands:

1. Click on **Analyze** from the menu bar.
2. Click on **Correlate** from the pull-down menu.
3. Click on **Bivariate** to open the Bivariate Correlations dialog box.
4. Click on the variable(s) that you wish to correlate, each followed by the **right arrow button** to move them into the Variables box.
5. Click on **OK**.

The output of both procedures appears in Figure 13.1.

The swarm of points in the scatter plot goes from lower left to upper right. We also see that there are no apparent outliers. In addition, the association appears linear, rather than (for instance) curvilinear.

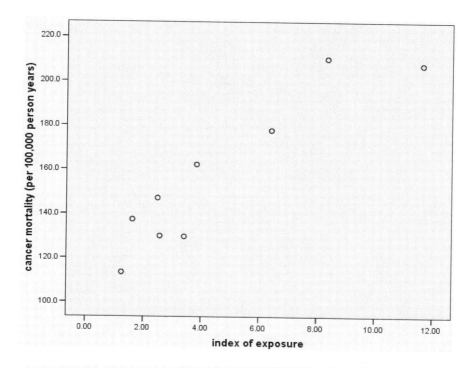

Correlations

		index of exposure	cancer mortality (per 100,000 person years)
index of exposure	Pearson Correlation	1	.926**
	Sig. (2-tailed)	.	.000
	N	9	9
cancer mortality (per 100,000 person years)	Pearson Correlation	.926**	1
	Sig. (2-tailed)	.000	.
	N	9	9

** Correlation is significant at the 0.01 level (2-tailed).

Figure 13.1 Scatter Plot and Correlation Coefficient of Exposure and Mortality

The correlation between exposure and mortality is +0.926, indicating that it is both positive and strong. Thus, higher levels of exposure to radioactive materials are strongly associated with higher levels of cancer mortality. The P value results from a test of significance (t-test of the hypothesis that the correlation is zero). The *t*-statistic for this case is not printed, but it is:

$$t = (0.926)\sqrt{\frac{(9-2)}{(1-0.926^2)}} = 6.507$$

The *P* value is obtained by referring this statistic to the t-distribution with 7 degrees of freedom. Here, because $P < .0005$, we conclude that the variables are significantly (positively) related.

13.2 SIMPLE LINEAR REGRESSION ANALYSIS

In linear regression analysis, sample data are used to estimate the intercept and slope of the "line of best fit" — the regression line — in a scatter plot. Simple regression analysis refers to a situation in which there is one independent and one dependent variable. The equation for the regression line in simple regression is $y = \alpha + \beta x$; where α and β are the *y*-intercept and slope, respectively. The slope β is usually of most interest because it tells the number of units increase (or decrease) in the dependent variable (*y*) associated with a one-unit increase in the independent variable (*x*).

We will illustrate this procedure using the "cancer.sav" data set to examine the association between mortality rates and exposure to radioactive materials.

After opening the "cancer.sav" data file, the steps for a regression analysis are:

1. Click on **Analyze** on the menu bar.
2. Click on **Regression** from the pull-down menu.
3. Click on **Linear** to open the Linear Regression dialog box (see Fig. 13.2).
4. Click on the variable that is your dependent variable ("mortalit"), and then click on the **top right arrow button** to move the variable name into the Dependent variable box.
5. Click on the variable that is your independent variable ("expose"), and then click on the **middle right arrow button** to move the variable name into the Independent(s) variable box.
6. Click on the **Statistics** button to open the Linear Regression: Statistics dialog box (Fig. 13.3).
7. By default, the **Estimates** and **Model fit** options are selected. Although we computed the correlation coefficient in a separate procedure, it is possible to calculate is as part of the regression procedure. To do so, select **Descriptives**. Another useful statistic is the confidence interval of the regression coefficients. To obtain this, click on **Confidence intervals**.

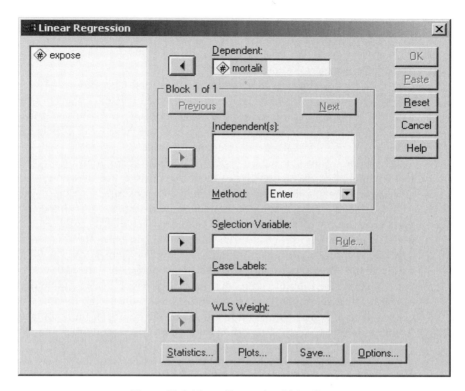

Figure 13.2 Linear Regression Dialog Box

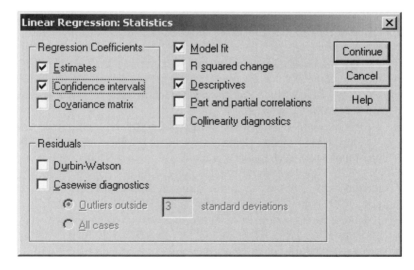

Figure 13.3 Linear Regression: Statistics Dialog Box

8.   Click on **Continue**.

9.   Click on **OK**.

The complete output is shown in Figure 13.4.

The first two tables — Descriptive Statistics and Correlations — are the result of selecting the **Descriptives** option. We see that the correlation between index of exposure and mortality rate is .926, exactly the same coefficient displayed in Figure 13.1. The only difference between these two tables is that the P value in Figure 13.4 is given as a one-tailed P value.

The square of the correlation ($0.926^2 = 0.858$) is the proportion of variation in y attributable to x; that is, 85.8% of the variation in cancer mortality is attributable to variation in radiation exposure. This is a very strong association.

The strength of association of the independent variable(s) with the dependent variable is also available in the Model Summary Table. This table represents the multiple correlation of the set of all independent variables (predictors) with the dependent variable. Because in simple regression there is only one predictor variable, the simple and multiple correlation coefficients are identical in number. However, the multiple correlation is always positive. The data analyst must remember that the multiple correlation does not indicate the direction of association! The R Square in this table represents the 85.8% of variation accounted for in mortality by index of exposure, as discussed previously.

Test of Significance for the Model

The output also includes a table labeled ANOVA (Analysis of Variance) located below the information about multiple correlation. This is a test of the significance of the model (that is, of the set of independent variables in predicting the dependent variable). The null hypothesis states that the set of all independent variables is not significantly related to the dependent variable. The Sig. column represents the P value for the test of significance of the model. In this case, $P < .0005$, so we conclude that the independent variable is significantly related to the dependent variable.

The other columns provide the detail and the building blocks from which the P value is determined. The sum of squares for Regression (8309.56) divided by the number of degrees of freedom (1) is the Mean Square for Regression (8309.56), which is the numerator of the F-ratio. The sum of squares labeled Residual (1373.95) is the sum of squared differences between the predicted values and the actual values of y, that is, the sum of squared deviations of the data around the regression line. These are combined to yield the proportion-of-variation statistic, r^2. The residual sum of squares divided by the number of degrees of freedom (7) is the variance of the residuals, 196.28 in the column labeled Mean Square. The square root of this value, 14.01, is the standard error of estimate $s_{y.x}$. The F-ratio is the ratio of these two mean squares,

Descriptive Statistics

	Mean	Std. Deviation	N
cancer mortality (per 100,000 person years)	157.344	34.7913	9
index of exposure	4.6178	3.49119	9

Correlations

		cancer mortality (per 100,000 person years)	index of exposure
Pearson Correlation	cancer mortality (per 100,000 person years)	1.000	.926
	index of exposure	.926	1.000
Sig. (1-tailed)	cancer mortality (per 100,000 person years)	.	.000
	index of exposure	.000	.
N	cancer mortality (per 100,000 person years)	9	9
	index of exposure	9	9

Variables Entered/Removed[b]

Model	Variables Entered	Variables Removed	Method
1	index of exposure[a]	.	Enter

[a]. All requested variables entered.

[b]. Dependent Variable: cancer mortality (per 100,000 person years)

Model Summary

Model	R	R Square	Adjusted R Square	Std. Error of the Estimate
1	.926[a]	.858	.838	14.0099

[a]. Predictors: (Constant), index of exposure

Figure 13.4 Regression Analysis Output with Descriptives and Confidence Intervals

ANOVA[b]

Model		Sum of Squares	df	Mean Square	F	Sig.
1	Regression	8309.556	1	8309.556	42.336	.000[a]
	Residual	1373.946	7	196.278		
	Total	9683.502	8			

a. Predictors: (Constant), index of exposure

b. Dependent Variable: cancer mortality (per 100,000 person years)

Coefficients[a]

Model		Unstandardized Coefficients		Standardized Coefficients	t	Sig.	95% Confidence Interval for B	
		B	Std. Error	Beta			Lower Bound	Upper Bound
1	(Constant)	114.716	8.046		14.258	.000	95.691	133.741
	index of exposure	9.231	1.419	.926	6.507	.000	5.877	12.586

a. Dependent Variable: cancer mortality (per 100,000 person years)

Figure 13.4 Regression Analysis Output with Descriptives and Confidence Intervals, *continued*

$$F = \frac{8309.556}{196.278} = 42.336 \cdot$$

The sum of squares are combined to obtain the proportion of variance in y explained by x, that is, the squared multiple correlation. In this example, this is

$$R^2 = \frac{8309.556}{9683.502} = .858 \cdot$$

Test of Significance for β

A test of the hypotheses H_0: $\beta = 0$ and H_1: $\beta \neq 0$ is given in the Coefficients table of the regression output. The t-statistic is $t = \frac{9.23}{1.418} = 6.507$. The P value, listed under Sig. in the output, is .000. Since this is smaller than most potential values of α (e.g., .05 or .01 or even .001), H_0 is rejected. We conclude that there is a nonzero (positive) association between exposure and cancer mortality in the population of counties represented by this sample.

At the outset of this study, researchers had reason to believe that a positive association might be found. Thus, a one-tailed test would have been appropriate with H_0: $\beta \leq 0$ and H_1: $\beta > 0$. The P value printed by SPSS is for a two-tailed test. To reject H_0 in favor of a one-sided alternative, $P/2$ must be less than α and the sign of the regression weight must be consistent with H_1. Both conditions are met in this example and H_0 is rejected in a one-tailed test as well.

The careful reader may notice that this t value and P are the same in the test of significance for β as they are for the test of the correlation coefficient (Section 13.1). When a study has just one numerical independent variable and one numerical dependent variable, the regression coefficient and the correlation coefficient have the same sign (+ or −) and the tests of significance are identical.

Estimating the Regression Equation

The least-squares estimates of the intercept and slope of the regression line are displayed in the Coefficients table of the output (Fig. 13.4) under the title Unstandardized Coefficients. Two values are listed in the column headed B; these are the intercept (a, 114.716) and the regression weight (b, 9.231), respectively. The equation of the least-squares line is thus $y = 114.716 + 9.231x$. (Instructions for having SPSS add the regression line to the scatter plot are given in a later section.)

SPSS also prints a form of β called the Standardized Coefficient, labeled Beta in the output. The standardized weight is the *number of standard deviations change in y associated with a one-standard deviation change in x*. Thus, in this example, a one-standard deviation increment in exposure is associated with a 0.93-standard deviation increment in cancer mortality — a large effect. When the units of x and y are familiar (e.g., income, time, body weight) the unstandardized ("raw") coefficient is easily interpreted. When scales are in less familiar units (e.g., psychological test scores) the standardized weight is a convenient way to express the relationship of x and y.

The Confidence Interval option produced two 95% intervals in the output, one for the slope and one for the intercept. The interval for the slope indicates that we are 95% confident that a one-unit increment in exposure is associated with an increase in mortality of at least 5.88 deaths per 100,000 person years and perhaps as much as 12.59 additional deaths. These values were obtained by adding to, and subtracting from, 9.23 a multiple of the standard error required to give the preselected confidence interval. In this example, the standard error is 1.42 (see Std. Error in Fig. 13.3), and the multiplier from the t-distribution with 7 degrees of freedom is 2.37.

Drawing the Regression Line

SPSS will draw the least-squares line (regression line) on a scatter plot. This is an option you may request as you create the plot or after doing so. For the latter instance, follow these steps:

1. Double click on the scatter plot chart to open the SPSS Chart Editor.

2. Select (click on) all the data points; they will become highlighted.
3. Click on **Elements** in the menu bar.
4. Select **Fit Line at Total** from the pull-down menu.
5. The Properties dialog box will open, with the option **Linear** fit line selected. Accept this default by clicking **Close**.
6. Click on the **X** in the upper right corner to close the Chart Edit Window.

The output is shown in Figure 13.5.

13.3   ANOTHER EXAMPLE: INVERSE ASSOCIATION OF X AND Y

As another example, open the data file "noise.sav," which has data on the relationship between the acceleration noise of an automobile and the speed for a section of highway. Make a scatter plot (with the regression line superimposed) with acceleration noise as the dependent variable (y) and speed (mph) as the independent variable (x), and perform a simple regression analysis (including the descriptive statistics). Your output from both procedures should look like Figure 13.6.

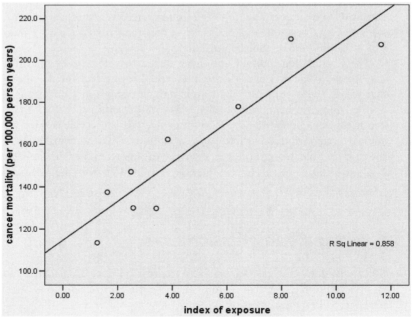

Figure 13.5 Scatter Plot with Regression Line

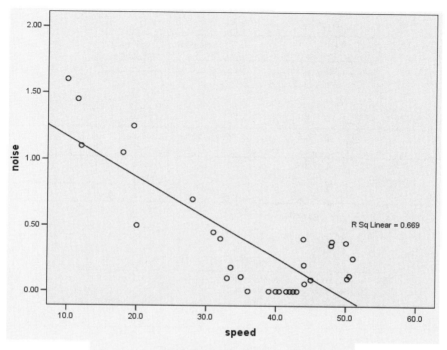

Descriptive Statistics

	Mean	Std. Deviation	N
NOISE	.3737	.46268	30
SPEED	36.087	12.3453	30

Correlations

		NOISE	SPEED
Pearson Correlation	NOISE	1.000	-.818
	SPEED	-.818	1.000
Sig. (1-tailed)	NOISE	.	.000
	SPEED	.000	.
N	NOISE	30	30
	SPEED	30	30

Variables Entered/Removed[b]

Model	Variables Entered	Variables Removed	Method
1	SPEED[a]	.	Enter

a. All requested variables entered.

b. Dependent Variable: NOISE

Figure 13.6 Scatter Plot and Regression Analysis of Noise and Speed

Model Summary

Model	R	R Square	Adjusted R Square	Std. Error of the Estimate
1	.818[a]	.669	.657	.27100

a. Predictors: (Constant), SPEED

ANOVA[b]

Model		Sum of Squares	df	Mean Square	F	Sig.
1	Regression	4.152	1	4.152	56.529	.000[a]
	Residual	2.056	28	.073		
	Total	6.208	29			

a. Predictors: (Constant), SPEED

b. Dependent Variable: NOISE

Coefficients[a]

Model		Unstandardized Coefficients		Standardized Coefficients	t	Sig.	95% Confidence Interval for B	
		B	Std. Error	Beta			Lower Bound	Upper Bound
1	(Constant)	1.480	.155		9.534	.000	1.162	1.798
	SPEED	-3.06E-02	.004	-.818	-7.519	.000	-.039	-.022

a. Dependent Variable: NOISE

Figure 13.6 Scatter Plot and Regression Analysis of Noise and Speed, *continued*

Notice that there is an inverse relationship between speed and acceleration noise. We first see this with the negative correlation (−0.818), indicating a strong, negative linear association. The inverse relationship is also indicated by the negative slope of the line in the scatter plot. The scatter plot also shows that there are no apparent outliers.

The test of whether the regression weight is significantly different from zero appears in the Coefficients table in the regression output. The sample regression weight is −0.0306. The *t*-statistic is $t = \dfrac{-0.0306}{0.004} = -7.519$ where 0.004 is the standard error of the regression coefficient. The *P* value given under the label Sig. is .000, which implies that $P < .0005$; noise is significantly inversely related to speed on the highway.

Given that we have a significant negative association of noise with speed, we ask about the strength of the relationship. Because speed is measured in familiar units (miles per hour, or mph), we may prefer to interpret the unstandardized regression weight (labeled B in the output). This tells us that every 1 mph increase in average speed on sections of the highway is associated with a .031-unit decrease in acceleration noise.

We have already seen that the correlation between speed and noise is strong and negative. In addition, the output produced by the regression analysis shows that the square of the correlation is 0.669. (Recall that the Multiple R is the absolute value of the correlation.) Thus, 66.9% of the variability in noise level is accounted for by speed and, by subtraction from 100%, 33.1% of variation in noise level is explained by other factors that were not included in this study.

No Relationship

Does it ever happen that a predictor variable is not related to the dependent variable? Yes, it does. This can be illustrated with the data in file "weather.sav" by examining the relationship between the amount of precipitation in an area and the temperature. Again, construct a scatter plot of the variables amount of precipitation ("precip") and temperature ("temp") (with the regression line) and compute the correlation coefficient as described in Section 13.1. The output is displayed in Figure 13.7.

The regression line appears to be rather flat and many of the points are far away from it. The correlation itself is small ($r = 0.124$) and nonsignificant ($P = .284$); less than 2% of the variability in "precipitation" is attributed to "temperature" ($0.124^2 = 0.015$). There is little to be gained by attempting to predict precipitation from temperature.

13.4   MULTIPLE REGRESSION ANALYSIS

The goal of multiple regression analysis is to explore how much variation in the dependent variable can be explained by variability in two or more independent variables. The equation of the regression line for two independent variables is: $y = \alpha + \beta_1 x_1 + \beta_2 x_2$. SPSS provides estimates and tests of the two most important parameters, β_1 and β_2, which determine the slope of the line; they reflect the "partial" contribution each independent variable has toward explaining the outcome (y).

Selecting the Order of Entry of the Independent Variables

When a regression analysis includes two or more independent variables, the order in which they are entered into the analysis is an important consideration. There are several approaches to entering the variables, depending on both the purpose of the research and philosophy of the researcher. In one approach, the researcher decides the order of entry according to the conceptual importance of the independent variables. This "hierarchical" procedure produces tests of the first independent variable, the *additional* contribution of the second independent variable, and so on. Other approaches use the strength of the partial correlations of the variables in the sample to determine which independent measures are included at each step. These procedures minimize the number of predictors in the final model. The examples in this chapter use the former (hierarchical) approach.

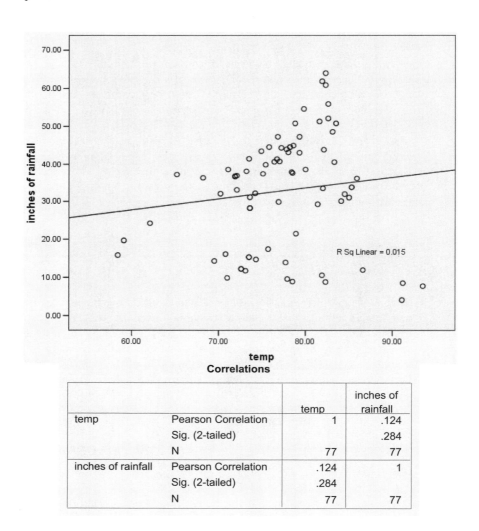

Correlations

		temp	inches of rainfall
temp	Pearson Correlation	1	.124
	Sig. (2-tailed)		.284
	N	77	77
inches of rainfall	Pearson Correlation	.124	1
	Sig. (2-tailed)	.284	
	N	77	77

Figure 13.7 Scatter Plot and Correlation Coefficient of Precipitation and Weather

We will illustrate the multiple regression procedure using the "cereal.sav" data set. Our interest is in examining the relationship between amount of fiber and carbohydrates in cereals and the number of calories per serving. Specifically, we wish to examine the additional effect of variations in fiber, over and above that of carbohydrate differences.

After opening the "cereal.sav" data file, do the following:

1. Click on **Analyze** on the menu bar.
2. Click on **Regression** from the pull-down menu.
3. Click on **Linear** to open the Linear Regression dialog box (see Fig. 13.2).
4. Click on the variable that is your dependent variable ("calories"), and then click on the **top right arrow button** to move the variable name into the Dependent variable box.
5. Click on the independent variable that you want to enter first into the model ("carbo"), and then click on the **second right arrow button** to move the variable name into the Independent(s) variable box.
6. Click on the **Next** button located above the Independent(s) box. The block should change to 2 of 2.
7. Click on the second independent variable ("fiber"), and then click on the **second right arrow button** to move the variable name into the Independent(s) variable box.
8. Click on Statistics to open the Linear Regression: Statistics dialog box (see Fig. 13.8).
9. Click on the **R squared change** and **Descriptives** (in addition to the default **Estimates** and **Model fit**).
10. Click on **Continue** to close the dialog box.
11. Click on **OK** to run the regression.

The complete output is shown in Figure 13.9.

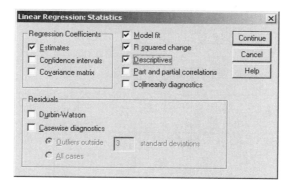

Figure 13.8 Linear Regression: Statistics Dialog Box

Descriptive Statistics

	Mean	Std. Deviation	N
CALORIES	106.8831	19.48412	77
CARBO	14.5974	4.27896	77
FIBER	2.1519	2.38336	77

Correlations

		CALORIES	CARBO	FIBER
Pearson Correlation	CALORIES	1.000	.251	-.293
	CARBO	.251	1.000	-.356
	FIBER	-.293	-.356	1.000
Sig. (1-tailed)	CALORIES	.	.014	.005
	CARBO	.014	.	.001
	FIBER	.005	.001	.
N	CALORIES	77	77	77
	CARBO	77	77	77
	FIBER	77	77	77

Variables Entered/Removed[b]

Model	Variables Entered	Variables Removed	Method
1	CARBO[a]	.	Enter
2	FIBER[a]	.	Enter

a. All requested variables entered.

b. Dependent Variable: CALORIES

Model Summary

Model	R	R Square	Adjusted R Square	Std. Error of the Estimate	R Square Change	F Change	df1	df2	Sig. F Change
1	.251[a]	.063	.050	18.98732	.063	5.029	1	75	.028
2	.333[b]	.111	.087	18.62206	.048	3.971	1	74	.050

a. Predictors: (Constant), CARBO

b. Predictors: (Constant), CARBO, FIBER

ANOVA[c]

Model		Sum of Squares	df	Mean Square	F	Sig.
1	Regression	1813.083	1	1813.083	5.029	.028[a]
	Residual	27038.865	75	360.518		
	Total	28851.948	76			
2	Regression	3190.153	2	1595.076	4.600	.013[b]
	Residual	25661.795	74	346.781		
	Total	28851.948	76			

a. Predictors: (Constant), CARBO

b. Predictors: (Constant), CARBO, FIBER

c. Dependent Variable: CALORIES

Figure 13.9 Multiple Regression Analysis Output

Coefficients[a]

Model		Unstandardized Coefficients		Standardized Coefficients		
		B	Std. Error	Beta	t	Sig.
1	(Constant)	90.221	7.739		11.658	.000
	CARBO	1.141	.509	.251	2.243	.028
2	(Constant)	99.867	9.002		11.094	.000
	CARBO	.762	.534	.167	1.427	.158
	FIBER	-1.911	.959	-.234	-1.993	.050

a. Dependent Variable: CALORIES

Excluded Variables[b]

Model		Beta In	t	Sig.	Partial Correlation	Collinearity Statistics Tolerance
1	FIBER	-.234[a]	-1.993	.050	-.226	.873

a. Predictors in the Model: (Constant), CARBO

b. Dependent Variable: CALORIES

Figure 13.9 *Continued*

Simple Correlations

The simple correlations between the independent and dependent variables are an important part of the results. These are shown in the Correlations table of the output. We see that there is a positive correlation between carbohydrates and calories (.251) and a negative correlation between fiber and calories (−.293). So, as the amount of carbohydrates in cereals increase, so does the number of calories. Conversely, cereals higher in fiber tend to have fewer calories. The output also gives P values for testing significance of each correlation. These are, however, simple correlations, and they do not include any consideration of the multiplicity of the independent variables.

The Full Model

The tests of significance for the full model are contained in the ANOVA table. The table lists two separate models because the variables were entered into the equation in separate steps. Model 1 corresponds to the simple model with calories regressed on carbohydrates. Model 2 represents the multiple regression with both carbohydrates and fiber as the independent variables. The *P* value is contained in the Sig. column. We see that *as a set*, the two predictors (carbohydrates and fiber) are significantly related to the number of calories in a cereal (*P* < .013).

The Model Summary table lists the multiple correlation for the two models. The squared correlation (R^2) indicated that carbohydrates alone account for 6.3% of variation in calories. For the full model, differences in fiber *and* carbohydrate content account for 11.1% of the variation in number of calories in cereals.

The Coefficients table lists the partial regression coefficients in raw and standardized form. From Model 2 we see that the regression equation is: CALORIES = 99.867 + .762 (CARBOHYDRATES) – 1.911 (FIBER). Each regression coefficient reflects the effect of the particular variable "above and beyond" (holding constant) the effect of the other independent variable(s). The coefficient for fiber is negative, indicating that independent of carbohydrates, higher fiber cereals have lower calories. That is, holding constant the amount of carbohydrates in a cereal, a one-gram increase in fiber content is associated with a decrease of 1.9 calories, on average.

The Sig. column contains the *P* values. Using a .05 significance level, the amount of fiber in a cereal is significantly related to calories, holding constant the amount of carbohydrates. Carbohydrates, on the other hand, are not related to calories, after controlling for fiber.

Incremental Models

On many occasions, the researcher is interested in examining the contribution of the independent variables to variation in the dependent variable in a particular order, that is, the first independent variable, the second above and beyond the first, and so on. An easy place to find these results is in the Model Summary table. It lists the proportion of variance accounted for (R Square Change), *F*-statistic (*F* Change), and *P* value (Sig. F Change) for each variable as it is entered. For instance, the squared multiple correlation for the first model (the model with carbohydrates as the only independent variable) is .063. (This is equal to the square of the simple correlation shown in the Correlations table.) The R^2 indicates that 6.3% of the variation in calories in cereals is explained by differences in the grams of carbohydrates. The *F* statistic (5.029) is significant at the .05 level (*P* < .028).

Model 2 shows that differences in fiber explain an *additional* 4.8% of the variation in calories in breakfast cereals (R Square Change). The additional explanatory power is significant at *P* < .050. Together, the two independent variables account for 11.1% of variation in calories (6.3% + 4.8%), or 100 times the squared multiple correlation for the full model.

Equivalently, the incremental effects may be found in the coefficients table. The regression weight for Model 1 (1.141) yields a t-statistic of 2.243 and P value of .028. The square of t (2.243^2) is 5.029, the F-statistic in the Model summary table; the P value is identical. The regression weight for FIBER in model 2 (–1.1911) yields a t-statistic of –1.993 and P value of .050. The square of t (1.993^2) is 3.971, the F-statistic in the Model Summary table; the P value is identical. The Model Summary table is the convenient place to find the changes in R^2 and tests for a hierarchical analysis. The Coefficients table is useful because it contains raw and standardized weights that also indicate the direction of the effects (positive or negative).

13.5   AN EXAMPLE WITH DUMMY CODING

Although regression requires numerical variables, it is possible to use categorical variables, especially dichotomous variables, as the independent variables. This is accomplished through a process called "dummy coding." Dummy coding for dichotomous variables involves recoding the variables into the values 0 and 1. If gender were the variable, for instance, females could be coded 0 and males 1 (or vice versa).

We illustrate this with the "bodytemp.sav" data file. This file contains information on body temperature, pulse rate, and sex (0 = female, 1 = male) for 130 adults. We perform a multiple regression analysis following the steps in Section 13.4 to determine whether pulse rate and gender are related to body temperature. We enter sex first, and pulse rate second. The output is given in Figure 13.10.

The simple correlations indicate that sex is negatively correlated with body temperature (r = –.198). Because sex is coded 0 = female and 1 = male, the negative correlation indicates that females have higher body temperature than do males (the correlation is statistically significant at the .05 level, with $P < .012$). The correlation matrix also indicates that pulse rate is positively correlated with body temperature (r = .254, $P < .002$).

The Model 2 of the ANOVA table indicates that together, the variables are significantly related to body temperature ($P < .001$). The change statistics (in the Model Summary table) show that sex is related to temperature ($P < .024$), and that it explains 3.9% of the variation in temperature. Pulse rate is also related to body temperature, after controlling for sex differences ($P < .005$). Pulse rate accounts for an additional 5.9% of variation in body temperature, over and above that explained by sex.

Descriptive Statistics

	Mean	Std. Deviation	N
body temperature (degrees Fahrenheit)	98.249	.7332	130
sex	.50	.502	130
pulse rate	73.76	7.062	130

Correlations

		body temperature (degrees Fahrenheit)	sex	pulse rate
Pearson Correlation	body temperature (degrees Fahrenheit)	1.000	-.198	.254
	sex	-.198	1.000	-.056
	pulse rate	.254	-.056	1.000
Sig. (1-tailed)	body temperature (degrees Fahrenheit)	.	.012	.002
	sex	.012	.	.264
	pulse rate	.002	.264	.
N	body temperature (degrees Fahrenheit)	130	130	130
	sex	130	130	130
	pulse rate	130	130	130

Variables Entered/Removed[b]

Model	Variables Entered	Variables Removed	Method
1	sex[a]	.	Enter
2	pulse rate[a]	.	Enter

[a.] All requested variables entered.

[b.] Dependent Variable: body temperature (degrees Fahrenheit)

Model Summary

Model	R	R Square	Adjusted R Square	Std. Error of the Estimate	Change Statistics R Square Change	F Change	df1	df2	Sig. F Change
1	.198[a]	.039	.032	.7215	.039	5.223	1	128	.024
2	.313[b]	.098	.084	.7017	.059	8.316	1	127	.005

[a.] Predictors: (Constant), sex

[b.] Predictors: (Constant), sex, pulse rate

Figure 13.10 SPSS Regression Output

ANOVA[c]

Model		Sum of Squares	df	Mean Square	F	Sig.
1	Regression	2.719	1	2.719	5.223	.024[a]
	Residual	66.626	128	.521		
	Total	69.345	129			
2	Regression	6.813	2	3.407	6.919	.001[b]
	Residual	62.532	127	.492		
	Total	69.345	129			

a. Predictors: (Constant), sex

b. Predictors: (Constant), sex, pulse rate

c. Dependent Variable: body temperature (degrees Fahrenheit)

Coefficients[a]

Model		Unstandardized Coefficients		Standardized Coefficients	t	Sig.
		B	Std. Error	Beta		
1	(Constant)	98.394	.089		1099.530	.000
	sex	-.289	.127	-.198	-2.285	.024
2	(Constant)	96.520	.656		147.240	.000
	sex	-.269	.123	-.184	-2.185	.031
	pulse rate	2.527E-02	.009	.243	2.884	.005

a. Dependent Variable: body temperature (degrees Fahrenheit)

Excluded Variables[b]

Model		Beta In	t	Sig.	Partial Correlation	Collinearity Statistics Tolerance
1	pulse rate	.243[a]	2.884	.005	.248	.997

a. Predictors in the Model: (Constant), sex

b. Dependent Variable: body temperature (degrees Fahrenheit)

Figure 13.10 *Continued*

Chapter Exercises

13.1 Using the "library.sav" data file:

a. Use SPSS to make a scatter plot of the variables "volumes" and "staff," and draw the least-squares line through the scatter plot.

b. Is the relationship between libraries' collection size and staff size positive or negative?

c. Judging from the scatter plot, would you estimate that the correlation is weak, moderate, or strong?

13.2 Using the data in "cereal.sav" use SPSS to:

   a. Perform a regression analysis to examine whether the sugar content in cereal is related to rating of taste.

   b. Compute the slope, the standard error of the slope, and a 95% confidence interval for the slope of the regression line.

   c. Is there a significant relationship between sugar content and rating? State the test statistic and P value and explain the nature of the relationship, if one exists.

   d. What is the value of the correlation between sugar content and rating?

   e. What is the square of the correlation and what is its interpretation?

13.3 Open the "fire.sav" data file and use the variables "sex" and "agility" to perform the following:

   a. Use SPSS to recode the sex variable so that males have a value of 1 and females have a value of 0. (Note: this is called "dummy coding" a variable.)

   b. Perform a regression analysis with the agility score as the dependent variable and sex, after recoding, as the independent variable.

   c. Is there a statistically significant relationship of agility with sex? What is the direction of the relationship? Be clear about which "sex" code is associated with better (lower) agility scores.

   d. Calculate the mean agility scores for males and for females and the difference between the two means. Compare this difference with the regression weight you obtained.

   e. Given the choice between the raw and standardized regression coefficient, which would you choose to emphasize in a written report of this study? Why?

13.4 Use the data in "sleep.sav" to perform a multiple regression analysis to examine whether the number of hours of dream sleep ("dream") is related to animals' likelihood of being preyed upon ("prey") and the exposure of the den during sleep ("sleepexp").

   a. Is the overall relationship of predation and den exposure associated with the amount of time spent dreaming?

   b. Are both predictors significant?

   c. What is the strength of the overall relationship?

d. Redo this analysis by first including the body weight variable. How does the overall relationship change with three predictors? After controlling for body weight, are the predation and den exposure indices related to dream time?

Chapter 14

ANOVA: Comparisons of Several Populations

In Chapter 11, we demonstrate the use of *t*-tests for comparing the means of two populations (such as males and females). The analysis of variance (ANOVA) allows us to compare the means of a larger number of populations (i.e., three or more). Through analysis of variance, variability among the group means is compared to the variability within the groups. If between-group variability is substantially greater than within-group variability, the means are declared to be significantly different. The ANOVA allows you to answer questions such as:

- Is political affiliation (Democrat, Republican, Independent) related to the number of elections in which people vote?
- Do three different techniques for memorizing words have different impacts on the mean number of words recalled?
- Is annual movie income from movie sales related to the type of movie (e.g., comedy, drama, horror, action, family)?
- Is heart rate after a step-workout affected by the height of the step (low or high) and the frequency of stepping (slow, medium, fast)?

In this chapter, we will use the One-Way ANOVA procedure in SPSS. We will also discuss techniques for obtaining follow-up tests to determine which group means differ from which others, and effect sizes to determine the magnitude of the differences between specific groups. We will also explore the Analysis of Variance of Ranks, which is appropriate when the assumptions required for performing ANOVA are violated. Finally, we will use the General Linear

Model (Univariate) procedure to conduct an ANOVA when there is more than one independent variable.

14.1   ONE-WAY ANALYSIS OF VARIANCE

We illustrate the ANOVA procedure for one independent variable by comparing the relationship between three ways of organizing information and children's ability to memorize words. The data are contained in the "words.sav" file. One of the two variables in this file denotes the organizational method provided to the child (1 = no information, 2 = words divided into three categories, 3 = words divided into six categories), and the other variable denotes the number of words memorized by the child. We shall use SPSS to test the null hypothesis that the mean number of words memorized by the three groups are equal, that is, H_0: $\mu_1 = \mu_2 = \mu_3$. In this example, the number of observations in the three conditions is equal (6).

Examining the Data

It is important to examine data visually prior to conducting an ANOVA test. One method for doing so is to visually inspect a scatter plot of the data. Another is by creating a box-and-whisker plot for the dependent variable (here, number of words memorized) separately for each level of the independent variable (here, the type of organizational method used). A third method available through SPSS is producing an error bar chart. Scatter plots are described in Chapters 5 and 13 and box-and-whisker plots in Chapter 5. Therefore, we will illustrate the error bar chart in this section. After opening the "words.sav" data file:

1. Click on **Graphs** from the menu bar.
2. Click on **Error bar** from the pull down menu.
3. Click on **Simple** and **Summaries for groups of cases** in the Error Bar dialog box.
4. Click on **Define** to open the Define Simple Error Bar: Summaries for Groups of Cases dialog box (Fig. 14.1).
5. Click on and move the dependent variable ("words") into the Variable box using the **top right arrow button.**
6. Click on and move the independent variable ("info_set") into the Category Axis variable box using the **bottom right arrow button.**
7. Select the representation for the bars in the Bars Represent box. You have option of selecting a specified confidence interval for the mean or a speci-

fied multiplier for the standard error of the mean. In this example, we maintain the default, which is a 95% confidence interval for the mean.

8. Click on **OK.**

The output is displayed in Figure 14.2.

The figure shows the 95% confidence interval for the average number of words memorized by children in each of the three information set groups. The circle represents the mean, and the horizontal lines the endpoints of the confidence interval. For instance, we see that children in the *no information* group memorized, on average, 4 words. Further, the 95% confidence interval for the mean is approximately 2.5 words to 5.5 words.

From inspection, it appears that the children in the *3 categories* group had the most success in memorizing words. To determine whether the groups differ significantly, however, we must conduct the one-way ANOVA test of significance.

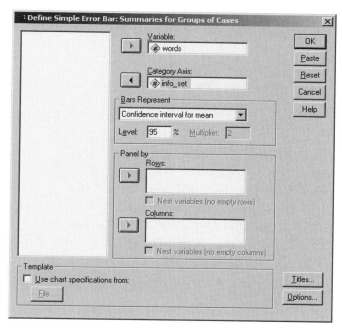

Figure 14.1 Define Simple Error Bar: Summaries for Groups of Cases Dialog Box

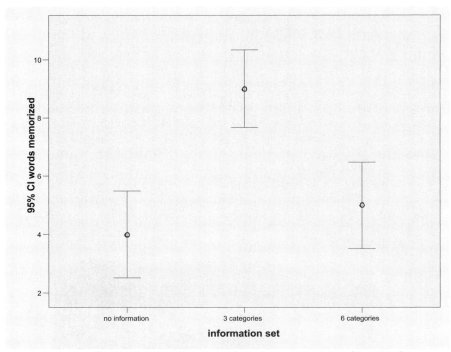

Figure 14.2 Error Bar Chart of Words Memorized by Information Set

Running the One-Way Procedure

To direct SPSS to perform the One-Way procedure:

1. Click on **Analyze** from the menu bar.
2. Click on **Compare Means** from the pull-down menu.
3. Click on **One-Way ANOVA** from the pull-down menu to open the One-Way ANOVA dialog box (Fig. 14.3).
4. Click on and move the "words" variable to the Dependent List box using the **top right arrow button.**
5. Click on and move the "info_set" variable to the Factor box using the **bottom right arrow button.**
6. Click on the **Options** button in the lower right corner of the One-Way ANOVA dialog box.
7. Click on the **Descriptives** option of the Statistics box. (This option provides mean, standard deviation, confidence interval, standard error, and the

minimum and maximum for each information category separately.) Also click on the **Means Plot** option to produce a graphical display of the means.

8. Click on **Continue** to close the One-way ANOVA: Option dialog box.

9. Click on **OK** to run the procedure.

The output containing descriptive statistics, the ANOVA table, and the means plot are displayed in Figure 14.4. The Sig. (.000) in the ANOVA table represents the P value corresponding to the F-ratio of 22.5 with 2 and 15 degrees of freedom. The null hypothesis that the population means are equal is rejected for any α greater than or equal to .0005. Thus, we conclude that there are differences among the three groups in the mean numbers of words memorized based on the information set.

From the Descriptives table and the Means plot, we see that the average number of words remembered by the no information group was 4; by the three categories group it was 9; and by the six categories group it was 5. SPSS also lists the range of words remembered for each group and computes a 95% confidence interval for each of the means. (This is similar to the information obtained in Section 14.1.)

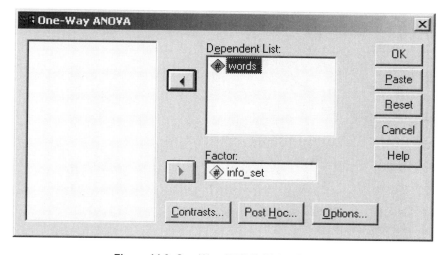

Figure 14.3 One-Way ANOVA Dialog Box

Descriptives

words memorized

	N	Mean	Std. Deviation	Std. Error	95% Confidence Interval for Mean		Minimum	Maximum
					Lower Bound	Upper Bound		
no information	6	4.0000	1.41421	.57735	2.5159	5.4841	2.00	6.00
3 categories	6	9.0000	1.26491	.51640	7.6726	10.3274	7.00	10.00
6 categories	6	5.0000	1.41421	.57735	3.5159	6.4841	3.00	7.00
Total	18	6.0000	2.56676	.60499	4.7236	7.2764	2.00	10.00

ANOVA

words memorized

	Sum of Squares	df	Mean Square	F	Sig.
Between Groups	84.000	2	42.000	22.500	.000
Within Groups	28.000	15	1.867		
Total	112.000	17			

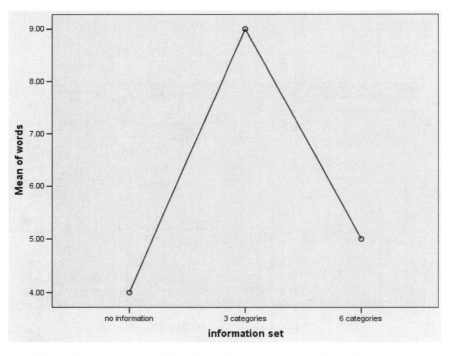

Figure 14.4 One-way ANOVA Listing with Descriptives and Means Plot

14.2   WHICH GROUPS DIFFER FROM WHICH, AND BY HOW MUCH?

Post-Hoc Comparisons of Specific Differences

When three or more populations are compared through analysis of variance, it is impossible to tell from the overall F-ratio which means differ significantly from which other means simply by inspection. You may wish to conduct tests for particular pairs of means. Conducting multiple tests inflates the probability of making a Type I error, however. There are several ways to compensate for this, many offered by SPSS. One approach is to use the Bonferroni correction procedure for computing the tests of significance for all possible pairs of information set (e.g., no information compared with three categories, no information compared with six categories, and three categories compared with six categories).

To illustrate, conduct the One-way procedure again, by following steps 1–8 as described in Section 14.1 under and then:

1.   Click on the **Post Hoc** button to open the One-Way ANOVA: Post Hoc Multiple Comparisons dialog box (Fig. 14.5).
2.   Click on the **Bonferroni** option; leave the significance level at the default .05.
3.   Click on **Continue** to close the dialog box.
4.   Click on **OK** to run the procedure.

The results should look similar to those displayed in Figure 14.4, with the addition of the Multiple Comparisons Table, contained in Figure 14.6. This table lists the results of tests of significance for all possible pairwise comparisons among the three information sets. For instance, the mean difference between number of words memorized by children in the no information group (I column) minus that of the 3 categories group (J column) was –5 words. Thus, the 3 categories group had a higher mean. (Note that the same information can be obtained in the pairing of no information in the J column and 3 categories in the I column.)

The Sig. column contains the P value for the test of significance. Here, $P < .0005$, indicating that the difference is statistically significant. The difference for no information minus 6 categories is –1, which is not statistically significant at the .05 level ($P < .673$). The difference between 3 categories minus 6 categories (4 words) is significant ($P < .0005$).

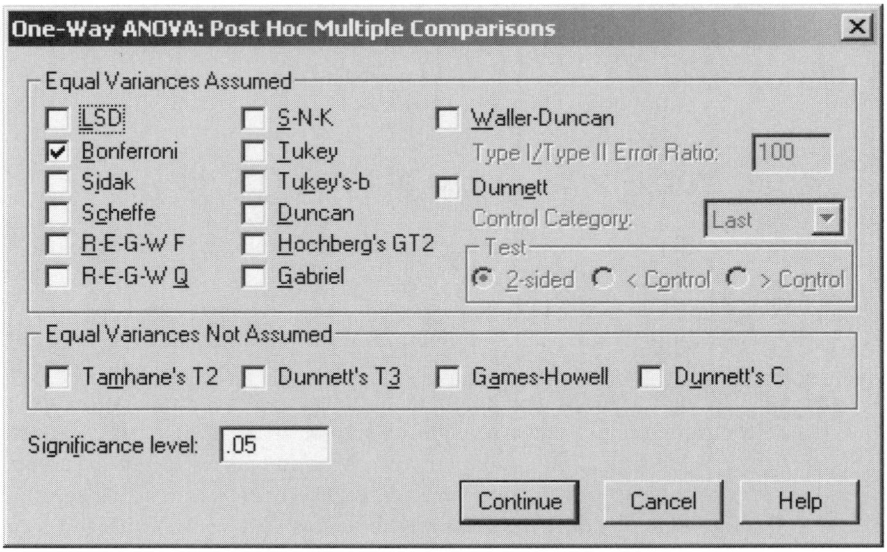

Figure 14.5 One-Way ANOVA: Post Hoc Multiple Comparisons Dialog Box

Multiple Comparisons

Dependent Variable: words memorized

Bonferroni

(I) information set	(J) information set	Mean Difference (I-J)	Std. Error	Sig.	95% Confidence Interval Lower Bound	Upper Bound
no information	3 categories	-5.0000*	.78881	.000	-7.1249	-2.8751
	6 categories	-1.0000	.78881	.673	-3.1249	1.1249
3 categories	no information	5.0000*	.78881	.000	2.8751	7.1249
	6 categories	4.0000*	.78881	.000	1.8751	6.1249
6 categories	no information	1.0000	.78881	.673	-1.1249	3.1249
	3 categories	-4.0000*	.78881	.000	-6.1249	-1.8751

*. The mean difference is significant at the .05 level.

Figure 14.6 Multiple Comparisons Table from the Post Hoc Option of the One-Way ANOVA

Therefore, we summarize the findings by stating that the 3 categories information set appears superior for word memorization to both the no information and the 6 categories sets, but that there is no difference between the no information and 6 categories sets. (Refer back to the graphs in Figure 4.2 and Figure 14.4; is our conclusion supported by the graphs?)

Effect Sizes

The *t*-tests allow us to determine which means differ significantly from which other means, but do not provide a clear indication of the *magnitude* of the differences. In some cases, the difference itself is a meaningful way to describe the strength of the effect. For example, students in the three categories condition remembered, on average, $9 - 4 = 5$ more words than those in the no information condition.

There are instances when the scale of the dependent variable is not familiar, as with many educational and psychological tests. In such situations, it is helpful to express the mean difference in standard deviation units; the result is called the effect size. For example, the effect size for the 5-word difference between children in the three-category group and those in the no information group is

$$\frac{9-4}{\sqrt{1.867}} = 3.66,$$

where the pooled within-group standard deviation is the square root of the mean square within group (found on the ANOVA table in Figure 14.4). The mean difference (3.66) is over 3 standard deviations, representing a very large effect size. On average, children given information grouped into three sets were 3.66 standard deviations above children in the no information group.

14.3 ANALYSIS OF VARIANCE OF RANKS

As with other statistical procedures, it is important to evaluate whether the data meet the assumptions underlying the ANOVA. These assumptions include (1) independence of observations, (2) a normal distribution of subgroup means, and (3) homogeneity of population variances.

While the analysis of variance is fairly robust with regard to violation of conditions (2) and (3), these assumptions should be examined routinely, especially if sample sizes are small to moderate. You can, for instance, inspect histograms of the dependent variable to assess departures from normality, as illustrated in Section 10.2. When there are more than two populations involved, the sample variances should also be examined to see if they are similar. In Figure 14.4, for example, the standard deviations of the three groups appear to be in the same general range. (There are also formal tests for equal variances, but they are not discussed in this manual.)

When the assumptions of normality and/or homogeneity of variance are severely violated, the ANOVA results may be misleading. In this case, an alternative procedure, the Kruskal-Wallis analysis of variance of ranks, is preferable.

The data file "bottle.sav" contains the daily output for three bottle capping machines. Using this data file, the procedure for using SPSS to perform the analysis of variance of ranks is:

1. Click on **Analyze** from the menu bar.
2. Click on **Nonparametric Tests** from the pull-down menu.
3. Click on **K Independent samples** from the pull-down menu. This opens the Tests for Several Independent Samples dialog box (Fig. 14.7).
4. Click on and move the "output" variable to the Test Variable List box using the **upper right arrow** button.
5. Click on and move the "machine" variable to the Grouping Variable box using the **lower right arrow** button.
6. Click on the **Define Range** button to open the Several Independent Samples: Define Range dialog box.
7. The machine variable is coded 1 through 3, so enter **1** in the minimum box and **3** in the maximum box.
8. Click on **Continue** to close the dialog box.
9. Notice that the Kruskal-Wallis H test is the default option in the Test Type box. Therefore, click on **OK** to run the procedure.

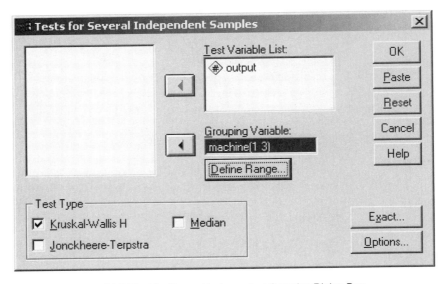

Figure 14.7 Test for Several Independent Samples Dialog Box

Ranks

	machine	N	Mean Rank
bottle cap output	Machine A	5	4.80
	Machine B	3	4.67
	Machine C	4	10.00
	Total	12	

Test Statistics[a,b]

	bottle cap output
Chi-Square	5.656
df	2
Asymp. Sig.	.059

a. Kruskal Wallis Test

b. Grouping Variable: machine

Figure 14.8 Kruskal-Wallis One-Way ANOVA Listing

The output from this test in contained in Figure 14.8. It shows the mean rank and sample size for each machine. For example, there are 4 machines of type C, and their mean rank is 10.00. The listing also reports the chi-square statistic (5.656) and the P value (.059). Using an "α level of .05, the null hypothesis is accepted and we conclude that the machines do not differ with respect to bottle cap output.

14.4 TWO-FACTOR ANALYSIS OF VARIANCE

In many instances, the data analyst will be interested in the relationship between two or more independent categorical variables and one dependent numerical variable. For instance, the "stepping.sav" data file contains information on an experiment conducted at The Ohio State University to explore the relationship between heart rate after a stepping exercise and the frequency of stepping (slow, medium, or fast) and the height of the step (low or high). Thirty individuals took part in the study.

Because there are two independent variables of interest, we conduct a two-factor or two-way (rather than a one-way) ANOVA. This allows us to look not only at the *main effect* of height and frequency, but also the *interaction* of the two variables. That is, we can ask if the relationship between frequency of stepping and heart rate after exercise is the same for different step heights.

We illustrate by opening the "stepping.sav" data file and doing the following:

1. Click on **Analyze** from the menu bar.

2. Click on **General Linear Model** from the pull-down menu and on **Univariate** from the supplementary pull-down menu. This will open the Univariate dialog box (Fig. 14.9).

3. Click on the variable "hr" (heart rate after exercise) and move it to the Dependent Variable box with the **top right arrow** button.

4. Click on "height" and move it to the Fixed Factor(s) box with the **second right arrow** button.

5. Repeat step 4 with the "frequenc" variable.

6. Click on the **Model** button to open the Univariate: Model dialog box. This is the box that allows you to construct your model. For instance, you can decide whether you want to include the interaction term (height-by-frequency) in the model. The default option, **Full factorial** includes both main effects and the interaction. We will maintain this option.

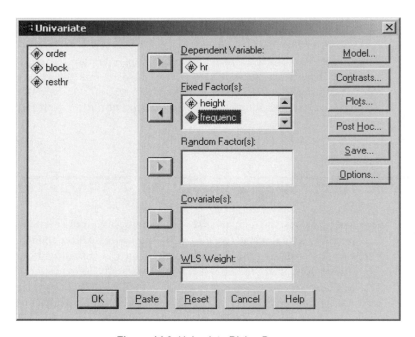

Figure 14.9 Univariate Dialog Box

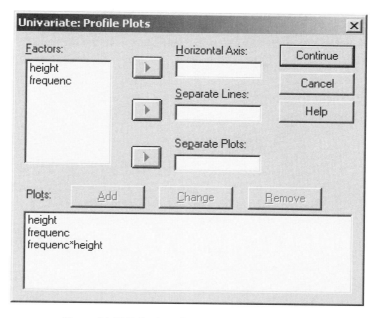

Figure 14.10 Univariate: Profile Plots Dialog Box

7. Click on **Continue.**

8. Click on **Plots** to open the Univariate: Profile Plots dialog box (Fig. 14.10). This option is similar to the means plots option in the one-way procedure. We can obtain one-way and two-way means graphs, however.

9. To obtain a graph of mean heart rate after exercise based on step height, click on the "height" variable and move it to the Horizontal Axis box using the **top right arrow** button.

10. Click on the **Add** button to move it into the Plots area.

11. Repeat steps 9-10 with the "frequenc" variable.

12. To obtain the interaction graph, click on and move the "frequenc" variable to the Horizontal Axis box with the **top right arrow** button and click on and move the "height" variable to the Separate Lines box with the **middle right arrow** button.

13. Click on **Add** to add this plot to the Plots area.

14. Click on **Continue** to close the dialog box.

15. Click on the **Options** button to open the Univariate: Options dialog box.

16. Click on the **Descriptive Statistics** option to produce means and standard deviation for the groups.

17. Click on **Continue**.

18. Click on **OK** to run the procedure.

The output of this procedure is displayed in Figure 14.11. The first table lists the between-subjects factors — the independent variables height of step and frequency of stepping. We see that there were 15 people in each of the height groups and 10 in each of the frequency groups.

The Descriptives table breaks down the groups, and displays the average heart rate after stepping for each cross-classification of the values of the independent variables. For instance, average heart rate after stepping for the 5 people in the low height, slow frequency group was 87.6; average in the high fast group was 136.8. This table also displays the marginal means — those by one variable. For instance, average heart rate for the 15 people in the low step group was 96.6, and 104.1 for the 10 people in the medium frequency group.

The ANOVA table is contained in the Tests of Between-Subjects Effects table. The Corrected Model row (with 5 degrees of freedom) refers to the full model — with two main effects and the interaction. The Sig. column displays the P value ($< .0005$), and indicates that model is significantly related to heart rate after stepping.

When there is an interaction effect in the model, it is important to interpret it before looking that the main effects. Thus, we examine the HEIGHT * FREQUENC effect first. The F-statistic is .540, and $P = .590$. We conclude that there is not a significant interaction between height and frequency of stepping on the heart rate. In other words, the effect of frequency of stepping on heart rate is the same whether the step height is low or high. (We can see this graphically in the last plot displayed in Figure 14.11. The blue line displays the relationship between heart rate and step frequency for low steps and the green line displays the relationship for high steps. The fact that the lines follow the same pattern and are almost parallel indicates the lack of a significant interaction.)

Because there is not a significant interaction, it is appropriate to interpret the main effects. The ANOVA table indicates that the height of the step is significantly related to heart rate ($F = 17.931$, $P < .0005$). In addition, frequency of stepping is also significantly related to heart rate ($F = 9.551$, $P < .0005$).

The descriptive statistics and the graphs indicate that heart rate increases with the height of the step (in the sample, mean is 96.6 for a low step and 118.2 for a high step). Because there are only two levels to this variable, we know where the significant difference lies.

Between-Subjects Factors

		Value Label	N
Height of step	.00	low	15
	1.00	high	15
Frequency of stepping	.00	slow	10
	1.00	medium	10
	2.00	fast	10

Descriptive Statistics

Dependent Variable: heart rate after stepping

Height of step	Frequency of stepping	Mean	Std. Deviation	N
low	slow	87.6000	9.09945	5
	medium	94.2000	8.89944	5
	fast	108.0000	16.70329	5
	Total	96.6000	14.26184	15
high	slow	103.8000	10.52141	5
	medium	114.0000	21.00000	5
	fast	136.8000	13.34916	5
	Total	118.2000	20.30904	15
Total	slow	95.7000	12.60555	10
	medium	104.1000	18.44180	10
	fast	122.4000	20.82306	10
	Total	107.4000	20.44437	30

Tests of Between-Subjects Effects

Dependent Variable: heart rate after stepping

Source	Type III Sum of Squares	df	Mean Square	F	Sig.
Corrected Model	7437.600[a]	5	1487.520	7.622	.000
Intercept	346042.800	1	346042.800	1773.214	.000
HEIGHT	3499.200	1	3499.200	17.931	.000
FREQUENC	3727.800	2	1863.900	9.551	.001
HEIGHT * FREQUENC	210.600	2	105.300	.540	.590
Error	4683.600	24	195.150		
Total	358164.000	30			
Corrected Total	12121.200	29			

a. R Squared = .614 (Adjusted R Squared = .533)

Figure 14.11 Results of Two-Way ANOVA

Estimated Marginal Means of heart rate after stepping

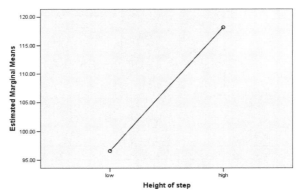

Estimated Marginal Means of heart rate after stepping

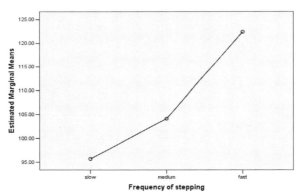

Estimated Marginal Means of heart rate after stepping

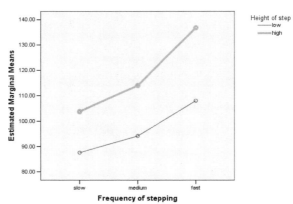

Figure 14.11 *Continued*

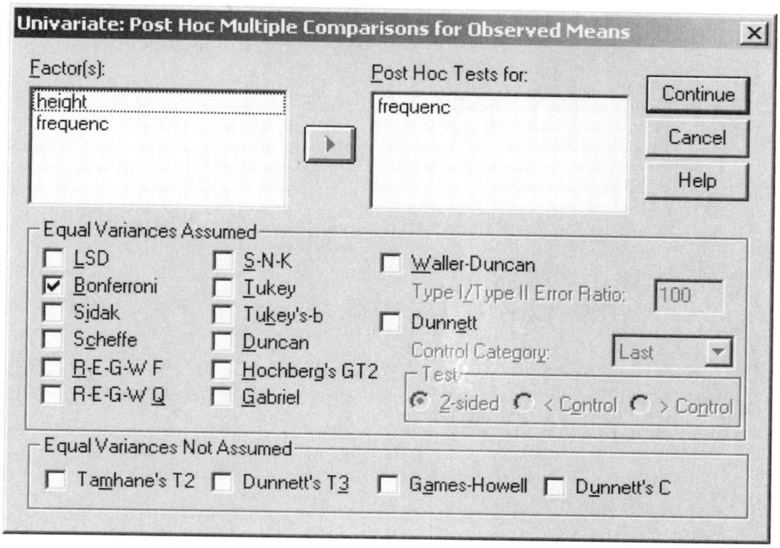

Figure 14.12 Univariate: Post Hoc Multiple Comparisons for Observed Means Dialog Box

The descriptive statistics and graphs also suggest that heart rate increases as frequency of stepping increases (the sample means are 95.7, 104.1, and 122.4 for slow, medium, and fast stepping, respectively). However, because there are three levels, we cannot say which levels differ statistically from one another without performing a follow-up test. To do so, re-run the ANOVA, adding the following steps before clicking **OK** to run the procedure.

1. From the Univariate dialog box, click on the **Post Hoc** button to open the Univariate: Post Hoc Multiple Comparisons for Observed Means dialog box (Fig. 14.12).

2. Click on and move the "frequenc" variable to the Post Hoc Tests for box with the **right arrow** button.

3. Click on the **Bonferroni** option.

4. Click on **Continue**.

The Multiple Comparisons table is listed in Figure 14.13. We read this table the same as in the one-way procedure. Results indicate that fast stepping is associated with a heart rate that is significantly higher than both slow stepping ($P =$ .001) and medium stepping ($P = .022$), but that there is no significant difference in heart rate between slow and medium stepping ($P = .574$).

Multiple Comparisons

Dependent Variable: heart rate after stepping
Bonferroni

(I) Frequency of stepping	(J) Frequency of stepping	Mean Difference (I-J)	Std. Error	Sig.	95% Confidence Interval Lower Bound	95% Confidence Interval Upper Bound
slow	medium	-8.4000	6.24740	.574	-24.4786	7.6786
	fast	-26.7000*	6.24740	.001	-42.7786	-10.6214
medium	slow	8.4000	6.24740	.574	-7.6786	24.4786
	fast	-18.3000*	6.24740	.022	-34.3786	-2.2214
fast	slow	26.7000*	6.24740	.001	10.6214	42.7786
	medium	18.3000*	6.24740	.022	2.2214	34.3786

Based on observed means.
    *· The mean difference is significant at the .05 level.

Figure 14.13 Bonferroni Multiple Comparisons of Frequency of Stepping

Chapter Exercises

14.1 Using the "hotdog.sav" data file:

    a. Test the hypothesis that there are no differences, on average, in calories based on type of hot dog. Use $\alpha = .05$ and state your conclusions in words.

    b. If you found a significant difference in (a), conduct the Bonferroni post hoc tests to determine where the differences lie. State your conclusions in one or two sentences.

    c. Compute an effect size for any of the comparisons that were significant in (b).

    d. Repeat steps (a) through (c) for sodium content.

14.2 The data file "movies.sav" contains information on number of weeks the movies for the year 2001 were in the Top 60.

    a. Perform an ANOVA to test whether weeks in Top 60 is independent of genre (type) of movie. Use $\alpha = .05$ and state your conclusions in words.

    b. If you found a significant difference in (a), conduct the Bonferroni post hoc tests to determine where the differences lie. State your conclusions in one or two sentences.

    c. Compute an effect size for any of the comparisons that were significant in (b).

Chapter 15

Exploratory Factor Analysis

Factor analysis is a data reduction technique. For instance, factor analysis can be used to identify the underlying components (factors) that explain the correlations among a set of variables. In this way, it is possible to employ a smaller set of measures (the factors) to explain a substantial portion of the total variance that is explained by all the original variables.

As an example, suppose you administer to a sample of adults a 25-item questionnaire asking questions about job satisfaction. Items may focus on rating of relationships with coworkers, intensity of interaction with coworkers, challenge provided by the daily tasks, valuing of the daily tasks, and the like. Factor analysis could be employed to reduce these individual variables to a much smaller set of factors that underlie job satisfaction such as satisfaction with coworkers and job commitment.

In exploratory factor analysis, there is no a priori assumption as to how the variables will combine to make factors. In confirmatory factor analysis, the researcher has a pre-established notion of which variables are associated with a given factor. In this chapter, we will use SPSS to conduct exploratory factor analysis.

15.1 CONDUCTING AN EXPLORATORY FACTOR ANALYSIS

We will examine exploratory factor analysis using data from a large study in which fourth grade teachers rated their students on a series of 15 behaviors (e.g., pays attention in class, thinks school is important, works well with others).[1] Each item was rated on a 5-point scale, with 1 representing "never" and 5 representing "always." We will conduct an exploratory factor analysis to find a few factors to explain most of the variation explained by the 15 behaviors. To do so, open the "behavior.sav" data file and follow the steps below.

1. Click on **Analyze** on the main menu bar.
2. Select **Data Reduction** from the pull-down menu.
3. Select **Factor** from the supplementary pull-down menu to open the Factor Analysis dialog box (see Fig. 15.1).
4. Highlight the variables to be included in the factor analysis (here, q1 through q15) and move them to the Variables box with the **top right arrow button**.
5. Click on **Extraction** to open the Factor Analysis: Extraction dialog box (Fig. 15.2). The default method of extraction is Principal components; the default criterion is to extract all factors with eigenvalues over 1; and the default is to analyze the correlation (rather than the covariance) matrix. Make any necessary changes to the settings and click **Continue.** (In this example, we will maintain the default settings.)
6. Click on **Rotation** to open the Factor Analysis: Rotation dialog box (Fig. 15.3). Select the rotation method desired (if any). In this example, we will select Varimax rotation (the most common form).[2] Click **Continue** to close the dialog box.
7. Click on **Options** to open the Factor Analysis: Options dialog box (Fig. 15.4). Click the **Sorted by size** option for the Coefficient Display Format section, and click on **Continue** to close the dialog box.
8. Click on **OK** to run the procedure.

---

[1] Finn, J. D., Pannozzo, G. M., & Voelkl, K. E. (1995). Disruptive and inattentive-withdrawn behavior and achievement among fourth-graders. Elementary School Journal , 95, 421-434.
[2] For a discussion of extraction and rotation, see Tabachnick, B. G., & Fidell, L. S. (1996). Using Multivariate Statistics. New York: HarperCollins.

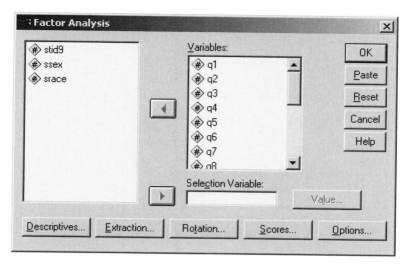

Figure 15.1 Factor Analysis Dialog Box

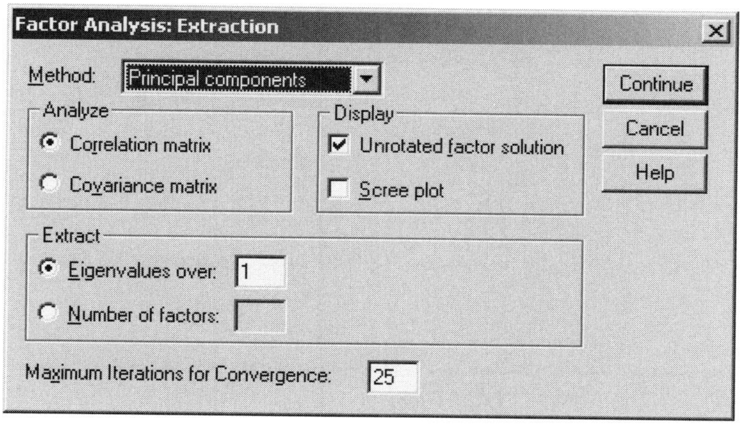

Figure 15.2 Factor Analysis: Extraction Dialog Box

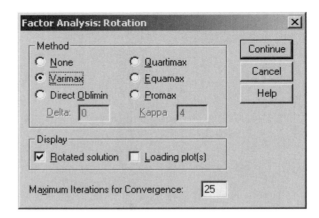

Figure 15.3 Factor Analysis: Rotation Dialog Box

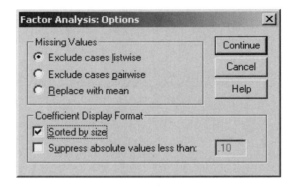

Figure 15.4 Factor Analysis: Options Dialog Box

15.2   INTERPRETING THE RESULTS OF THE FACTOR ANALYSIS PROCEDURE

The output of the factor analysis procedure we completed in Section 15.1 should appear as in Figure 15.5. The first table of the output reports communalities, which indicate the amount of variance in each variable that is accounted for. Because we have selected principal components analysis as the extraction method and opted to analyze the correlation matrix, the initial communalities for each variable are always 1.000. The extraction column indicates the variance in each variable accounted for by the factors (components) extracted as a result of the factor analysis.

Communalities

	Initial	Extraction
PAYS ATTENTION IN CLASS	1.000	.731
COMPLETES HOMEWORK ON TIME	1.000	.786
WORKS WELL W OTHERS	1.000	.610
TRIES TO DO WORK WELL	1.000	.749
PARTICIPATES IN DISCUSSIONS	1.000	.654
COMPLETES SEAT WORK	1.000	.780
THINKS SCHOOL IS IMPORTANT	1.000	.741
DOES EXTRA WORK	1.000	.644
MAKES EFFORT	1.000	.777
ASKS QUESTIONS	1.000	.699
TRIES TO FINISH DIFFICULT WORK	1.000	.742
RAISES HAND TO TALK	1.000	.613
SEEKS REFERENCE MATERIAL	1.000	.621
DISCUSSES OUTSIDE OF CLASS	1.000	.652
ATTENDS EXTRACURRICULAR ACTIVITIES	1.000	.332

Extraction Method: Principal Component Analysis.

Total Variance Explained

Component	Initial Eigenvalues Total	% of Variance	Cumulative %	Extraction Sums of Squared Loadings Total	% of Variance	Cumulative %	Rotation Sums of Squared Loadings Total	% of Variance	Cumulative %
1	8.724	58.157	58.157	8.724	58.157	58.157	6.223	41.484	41.484
2	1.407	9.382	67.539	1.407	9.382	67.539	3.908	26.055	67.539
3	.793	5.285	72.824						
4	.563	3.755	76.578						
5	.506	3.376	79.954						
6	.476	3.171	83.125						
7	.382	2.548	85.673						
8	.329	2.192	87.865						
9	.322	2.146	90.012						
10	.307	2.048	92.060						
11	.290	1.934	93.994						
12	.263	1.756	95.750						
13	.248	1.650	97.401						
14	.220	1.464	98.864						
15	.170	1.136	100.000						

Extraction Method: Principal Component Analysis.

Figure 15.5 Output for Factor Analysis with Behavior Variables

The second table reports information on Initial Eigenvalues, Extraction Sums of Squared Loadings, and (if appropriate) Rotation Sum of Squared Loadings. Eigenvalues are measures of the variance in all the variables that are accounted for by a given factor. In the initial solution, there are as many factors (components) as there are variables. In our example, the first component accounts for 8.724 of the total variance; the second, 1.407, etc. The sum of the individual eigenvalues is equivalent to the variance in all of the variables (here, 15). The "% of variance" column represents the ratio of the variance explained by the individual component to the total variance. In our example, the first component accounts for 58.157% of the total variance (8.724   15 = 58.157%). The initial solution is the complete solution; it explains 100% of the variance in the original variables. (This can be seen in the cumulative percent column; note that the cumulative percent for the fifteenth component is 100.00%.)

Component Matrix[a]

	Component	
	1	2
MAKES EFFORT	.869	-.149
THINKS SCHOOL IS IMPORTANT	.847	-.152
TRIES TO DO WORK WELL	.835	-.227
COMPLETES SEAT WORK	.832	-.297
TRIES TO FINISH DIFFICULT WORK	.831	-.228
PAYS ATTENTION IN CLASS	.831	-.203
COMPLETES HOMEWORK ON TIME	.829	-.315
DOES EXTRA WORK	.787	.155
RAISES HAND TO TALK	.750	.225
SEEKS REFERENCE MATERIAL	.738	.276
ASKS QUESTIONS	.737	.395
PARTICIPATES IN DISCUSSIONS	.729	.351
WORKS WELL W OTHERS	.720	-.303
ATTENDS EXTRACURRICULAR ACTIVITIES	.411	.404
DISCUSSES OUTSIDE OF CLASS	.555	.586

Extraction Method: Principal Component Analysis.
[a.] 2 components extracted.

Rotated Component Matrix[a]

	Component	
	1	2
COMPLETES HOMEWORK ON TIME	.856	.229
COMPLETES SEAT WORK	.848	.246
TRIES TO DO WORK WELL	.810	.304
TRIES TO FINISH DIFFICULT WORK	.807	.301
PAYS ATTENTION IN CLASS	.793	.321
MAKES EFFORT	.792	.387
THINKS SCHOOL IS IMPORTANT	.776	.372
WORKS WELL W OTHERS	.762	.175
DISCUSSES OUTSIDE OF CLASS	.107	.800
ASKS QUESTIONS	.367	.752
PARTICIPATES IN DISCUSSIONS	.386	.711
SEEKS REFERENCE MATERIAL	.438	.655
RAISES HAND TO TALK	.477	.621
DOES EXTRA WORK	.548	.586
ATTENDS EXTRACURRICULAR ACTIVITIES	9.697E-02	.568

Extraction Method: Principal Component Analysis.
Rotation Method: Varimax with Kaiser Normalization.
[a.] Rotation converged in 3 iterations.

Figure 15.5 Output for Factor Analysis with Behavior Variables, *continued*

One of the goals of factor analysis is parsimony, so we must balance explaining a substantial percentage of variation with limiting the number of factors (components) selected. The Extraction Sum of Squared Loadings section displays information regarding the factors/components extracted. Here, SPSS has extracted two factors/components, based on the extraction criterion we chose —

eigenvalues greater than 1. These two components explain 67.539% of the variance in the original variables. Because we selected principal components analysis as the method of extraction, the "Total," "% of Variance," and "Cumulative %" columns for this section are identical to those of the first two components in the Initial Eigenvalues section.

Because we selected varimax rotation, there is a third section — "Rotation Sum of Squared Loadings." The variance accounted for by individual rotated factors/components may differ from that accounted for by individual unrotated factors/components. However, the final cumulative percent of variance accounted for is equivalent (here, 67.539%).

The next two tables, the Component Matrix and the Rotated Component Matrix represent factor loadings for each of the original variables on the original (unrotated) and rotated solution, respectively. For the unrotated solution, factor loadings are the correlations between the specific item and unrotated factor. For the rotated solution, factor loadings are the partial correlation between the item and the rotated factor. These correlations are helpful in determining the structure underlying the individual items. That is, you can examine variables (here, responses to the questionnaire items) that have high loadings (correlations or partial correlations) on a particular factor, and attempt to discern a common thread among them.

The correlations (loadings) from the rotated solution are sometimes easier to interpret for this purpose. For instance, varimax rotation tends to simplify the factors. That is, items tend to load heavily on only one factor. Thus, interpretation is simplified because it is easier to determine which variable is associated with which factor. In addition, because we selected "Sorted by size" (see step 7 above), the variables are already somewhat grouped. That is, the first 8 questionnaire items (beginning with "completes homework on time" and ending with "works well with others") load heavily on the first component, and the remaining 7 items ("discusses outside of class" through "attends extracurricular activities") load heavily on the second component.

The final task is to determine what the items within the task have in common. The first component seems to relate to academic work and behavior, and the second component seems to relate to social interactions and initiative.

15.3   SCALE RELIABILITY

An important characteristic of educational and psychological measurements is their reliability. Reliability indicates the amount of variation to expect in the measurement from one occasion to another. Indexes of reliability — reliability coefficients — can be computed that range from 0.0 (the measure has no reliability and may be expected to vary a great deal) to 1.0 (the measure has perfect reliability and will be consistent from one occasion to another. This is the scale of all the reliability coefficients discussed here.

The reliability of a measure can be determined in several ways. One way is to administer the measurement instrument at two different points in time — preferably close together — and to correlate the results from the two administrations. This is called "test-retest reliability" and can be done with SPSS using the correlation procedure described in Chapter 5. This approach can be taken with any measurement instrument having one or more items.

For multi-item measurement scales, more common indexes of reliability reflect the "internal consistency" of the scale. To compute internal consistency coefficients, the measure only needs to be administered once. The indexes reflect the extent to which scores are consistent from one portion of the measurement instrument to another. One internal consistency measure — the "split half" coefficient — is obtained by scoring two halves of the measurement scales and then correlating the results on one half with the results on the other half. The halves may be the first half of the instrument and the second half or, preferably, the odd-numbers items (1, 3, 5, ...) and the even-numbered items (2, 4, 6, ...). The correlation between scores on the two halves, however, only gives an estimate of the reliability of half of the test. It must be adjusted to estimate the reliability of the full-length test. The Spearman-Brown prophecy formula accomplishes this; it is applied automatically by SPSS when the split-half method is chosen.

A commonly used coefficient of internal consistency — Cronbach's coefficient alpha — is based on the consistency of responses from one item to another. It is often preferable to split-half reliability because it avoids making an arbitrary decision about what the halves of the measurement instrument are.

Coefficient alpha is interpreted in the same way as other reliability coefficients, with 0.0 indicating that the instrument has no reliability and 1.0 indicating the highest possible reliability. When you use the Reliability Analysis procedure in SPSS, the output includes item-by-item statistics indicating how each item contributes to the internal consistency of the measuring instrument. Items that do not contribute much to the reliability of the total measurement can be rewritten or deleted to improve the measurement scale. Also, the output indicates what the reliability of the total measurement scale would be if all items were "standardized" to the same metric (item mean = 0.0; item standard deviation = 1.0). If this value is much higher than the alpha-value for unstandardized items, it may be worthwhile taking this step when scoring the total scale for further use.

WARNING: SPSS only computes coefficient alpha correctly when all items are scored in the same direction. Thus, if high scores indicate high levels of the construct on some items and low levels of the construct on other items, one or the other set of items must be reverse scored. For example, suppose that you are rating student behavior and one item is "student studies hard" with response categories 1 = never, 2 = sometimes, 3 = always. Perhaps another item is "student is disruptive" with response categories 1 = never, 2 = sometimes, 3 = always. In order for coefficient alpha to be computed correctly by SPSS, one or the other item needs to be reversed. You can accomplish this by recoding the second item (recode "1" to "3" and "3" to "1") or by a transformation (scores on the transformed variable are "4 - the original item score") (see Chapter 1). In either case the revised item scores are used in the Reliability analysis.

15.4   COMPUTING SPLIT-HALF COEFFICIENT ESTIMATES

Recall that split-half coefficients represent the consistency in responding between the first half and second half of the items listed on the reliability analysis. Choose the items to include in each half so that the halves are as equivalent as possible. A common method of splitting the items is to assign the odd numbered items to one half, and the even numbered items to the other half.

Let's calculate the internal consistency of the behavior scale described in Section 15.1. To compute a split-half coefficient:

1. Click on **Analyze** on the main menu bar.
2. Click on **Scale** from the pull-down menu.
3. Click on **Reliability Analysis** to open the Reliability Analysis dialog box (see Fig. 15.6).
4. Highlight the odd numbered variables (q1, q3, q5 to q15) to be included in the first half of the reliability analysis and move them into the Items box with the **right arrow button**.
5. Highlight the even numbered variables (q2, q4, q6 to q14) to be included in the second half of the reliability analysis and move them into the Items box with the **right arrow button**.
6. Click on **Split-half** in the drop down menu labeled Model.
7. Click on **OK** to run the procedure.

The results of the analysis are shown in Figure 15.7. The reliability coefficient is calculated using the Spearman-Brown formula. Because there was an odd number of items in the scale, the split produced an unequal number of items in each half, so the unequal-length Spearman-Brown reliability coefficient is reported. Notice that the value of the coefficient is .956 indicating that the scale has good internal consistency. Additional statistics such as descriptives and inter-item correlations may be obtained by clicking on the **Statistics** button in the reliability analysis dialog box.

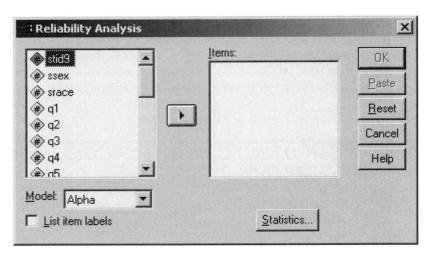

Figure 15.6 Reliability Analysis Dialog Box

Reliability Statistics

Cronbach's Alpha	Part 1	Value	.896
		N of Items	8[a]
	Part 2	Value	.890
		N of Items	7[b]
	Total N of Items		15
Correlation Between Forms			.916
Spearman-Brown	Equal Length		.956
Coefficient	Unequal Length		.956
Guttman Split-Half Coefficient			.954

[a.] The items are: PAYS ATTENTION IN CLASS, WORKS WELL W OTHERS, PARTICIPATES IN DISCUSSIONS, THINKS SCHOOL IS IMPORTANT, MAKES EFFORT, TRIES TO FINISH DIFFICULT WORK, SEEKS REFERENCE MATERIAL, ATTENDS EXTRACURRICULAR ACTIVITIES.

[b.] The items are: ATTENDS EXTRACURRICULAR ACTIVITIES, COMPLETES HOMEWORK ON TIME, TRIES TO DO WORK WELL, COMPLETES SEAT WORK, DOES EXTRA WORK, ASKS QUESTIONS, RAISES HAND TO TALK, DISCUSSES OUTSIDE OF CLASS.

Figure 15.7 Output for Split-Half Reliability Analysis

15.5 COMPUTING CRONBACH ALPHA COEFFICIENT

Cronbach alpha is another coefficient that can be used to determine reliability of a scale. To compute coefficient alpha for the 15-item behavior scale:

1. Click on **Analyze** on the main menu bar.

Reliability Statistics

	Cronbach's Alpha	Cronbach's Alpha Based on Standardized Items	N of Items
	.945	.946	15

Item-Total Statistics

	Scale Mean if Item Deleted	Scale Variance if Item Deleted	Corrected Item-Total Correlation	Squared Multiple Correlation	Cronbach's Alpha if Item Deleted
PAYS ATTENTION IN CLASS	50.35	141.862	.786	.673	.940
COMPLETES HOMEWORK ON TIME	50.17	140.249	.778	.746	.940
WORKS WELL W OTHERS	50.18	144.029	.661	.535	.943
TRIES TO DO WORK WELL	50.40	138.176	.787	.694	.940
PARTICIPATES IN DISCUSSIONS	50.56	140.602	.697	.578	.942
COMPLETES SEAT WORK	50.04	141.090	.782	.749	.940
THINKS SCHOOL IS IMPORTANT	50.16	139.077	.807	.693	.939
DOES EXTRA WORK	51.49	137.128	.755	.608	.940
MAKES EFFORT	50.38	138.141	.830	.735	.939
ASKS QUESTIONS	50.76	141.420	.710	.599	.941
TRIES TO FINISH DIFFICULT WORK	50.32	138.041	.782	.696	.940
RAISES HAND TO TALK	50.38	140.880	.716	.568	.941
SEEKS REFERENCE MATERIAL	51.17	138.807	.705	.561	.942
DISCUSSES OUTSIDE OF CLASS	51.31	144.055	.524	.413	.946
ATTENDS EXTRACURRICULAR ACTIVITIES	50.75	150.259	.379	.189	.949

Figure 15.8 Output for Coefficient Alpha Reliability Analysis

2. Select **Scale** from the pull-down menu.

3. Click on **Reliability Analysis** to open the Reliability Analysis dialog box.

4. Highlight the variables (q1 to q15) to be included in the reliability analysis and move them into the Items box with the **right arrow button**.

5. Click on **Statistics** to open the Reliability Analysis: Statistics dialog box.

6. Click on **Item** and **Scale** and **Scale if item deleted** in the Descriptives for area.

7. Click on **Correlations** in the Inter-item area.

8. Click on **Continue**.

9. Click on **Alpha** in the drop down menu labeled Model.

10. Click on **OK** to run the procedure.

A portion of the output is shown in Figure 15.8. The Reliability Statistics Table indicates that Cronbach's alpha for the 15-item scale is .945, indicating good reliability. The Item-Total Statistics table displays information about the scale as if it were calculated without each item. This allows the analyst to gain some information on how individual items contribute to the whole.

For instance, the last column indicates the value of Cronbach's alpha if each single item is deleted. In this example, omitting the item that measures the extent to which the student "thinks school is important" would result in a reliability coefficient of .939. This is a decrease from the calculation with all items (.945). Thus, we consider this item — thinks school is important — to contribute positively to the construct the scale measures. On the other hand, if the reliability were calculated without reference to the extent to which the student "attends extracurricular activities," the reliability would increase slightly, to .949.

Chapter Exercises

15.1 The "crime.sav" data file contains information on crime rates per 100,000 in the 50 states in the US. Use this datafile to conduct an exploratory factor on US crime rates, with the default extraction methods and criterion, and selecting varimax rotation. On the basis of the factor analysis, answer the following:

    a. What percent of total variance is accounted for by the component(s)?

    b. What is the number of eigenvalues greater than 1?

    c. How many components are extracted?

    d. If there is more than one component, use the rotated component matrix to interpret the components. What meaning would you attach to the factor(s)?

15.2 Based on the results of the factor analysis of classroom behaviors conducted in this chapter, conduct a reliability analysis using the eight variables that loaded heavily on the first factor, academic work. These variables are: Q1, Q2, Q3, Q4, Q6, Q7, Q9, Q11.

   a. What is the coefficient alpha?

   b. Does the reliability of the academic scale increase or decrease if you exclude the item "completes homework on time" (Q11)? Hint: use the options in the Reliability Analysis: Statistics dialog box.

15.3 Using the "fire.sav" data file, calculate the reliability coefficient for a fitness scale based on the items: stair clime time, body drag time, and obstacle course time. Discuss the internal consistency of the fitness scale.

Data Files

The data files used in this manual are contained on the Internet. Many are available through the Spring-Verlag web site, in SPSS ".sav" format. The remaining files are available in raw data format on the websites provided; consult Chapter 1 for reading raw data into SPSS. **Note: In the text of this book, we assume that you have retrieved all data, and saved them as SPSS ".sav" files. Thus, all datfiles are referred to as <filename>.sav.**

BEHAVIOR Teacher ratings of 4th grade students' behavior ($n = 2217$).

Variable Name	Variable Description
stid9	Student identification number
ssex	Student sex
srace	Student race
1 = white	
2 = black	
3 = Asian	
4 = Hispanic	
5 = Native American	
6 = other	
q1	Pays attention in class
1 = never	
2 = occasionally	
3 = sometimes	
4 = usually	
5 = always	

q2   Completes homework on time
- 1 = never
- 2 = occasionally
- 3 = sometimes
- 4 = usually
- 5 = always

q3   Works well with others
- 1 = never
- 2 = occasionally
- 3 = sometimes
- 4 = usually
- 5 = always

q4   Tries to do work well
- 1 = never
- 2 = occasionally
- 3 = sometimes
- 4 = usually
- 5 = always

q5   Participates in discussions
- 1 = never
- 2 = occasionally
- 3 = sometimes
- 4 = usually
- 5 = always

q6   Completes seat work
- 1 = never
- 2 = occasionally
- 3 = sometimes
- 4 = usually
- 5 = always

q7   Thinks school is important
- 1 = never
- 2 = occasionally
- 3 = sometimes
- 4 = usually
- 5 = always

q8   Does extra work
- 1 = never
- 2 = occasionally
- 3 = sometimes
- 4 = usually
- 5 = always

q9   Makes effort
- 1 = never

      2 = occasionally
      3 = sometimes
      4 = usually
      5 = always
q10   Asks questions
      1 = never
      2 = occasionally
      3 = sometimes
      4 = usually
      5 = always
q11   Tries to finish difficult work
      1 = never
      2 = occasionally
      3 = sometimes
      4 = usually
      5 = always
q12   Raises hand to talk
      1 = never
      2 = occasionally
      3 = sometimes
      4 = usually
      5 = always
q13   Seeks reference material
      1 = never
      2 = occasionally
      3 = sometimes
      4 = usually
      5 = always
q14   Discusses outside of class
      1 = never
      2 = occasionally
      3 = sometimes
      4 = usually
      5 = always
q15   Attends extracurricular activities
      1 = never
      2 = occasionally
      3 = sometimes
      4 = usually
      5 = always

BODYTEMP Body temperature and pulse rate for adults (n = 130). Available through *Journal of Statistics Education* Data Archive: http://www.amstat.org/publications/jse/datasets/normtemp.dat

Variable Name	**Variable Description**
temp	Body temperature, degrees Fahrenheit
sex	Sex
0 = female	
1 = male	
pulse	Pulse rate

BOTTLE Daily output of 12 bottle capping machines (n = 12). Kruskal, W.H. & Wallis, W.A. (1952). Use of ranks in one-criterion analysis of variance. *Journal of the American Statistical Association, 47,*583–621.

Variable Name	**Variable Description**
Machine	Machine identification code
Output	Number of bottles capped

CANCER Exposure to radioactive materials and cancer mortality rate (n = 9). Fadeley, R.C. (1965). Oregon malignancy pattern physiographically related to Hanford, Washington, radioisotope storage. *Journal of Environmental Health, 27,* 883–897.

Variable Name	**Variable Description**
expose	Index of exposure
mortalit	Cancer mortality (per 100,000 person years)

CARS Age, color, and owner of cars parked in university parking lot (n = 64). Collected in the parking lot of a northeastern university.

Variable Name	**Variable Description**
color	Color of car
1 = blue	
2 = gray	
3 = red	
4 = black	
5 = green	
6 = white	
7 = brown	
age	Age of car (years)
owner	Car owner
1 = faculty/staff	
2 = student	

CEREAL Nutritional information for breakfast cereals (*n* = 77) Available through the Data and Story Library: http://lib.stat.cmu.edu/DASL/Datafiles/Cereals.html

Variable Name	**Variable Description**
name	Name of cereal
manufac	Manufacturer
1 = American Home Foods	
2 = General Mills	
3 = Kellogg's	
4 = Nabisco	
5 = Post	
6 = Quaker Oats	
7 = Ralston Purina	
type	Type of cereal
1 = cold	
2 = hot	
calories	Calories per serving
protein	Protein grams
fat	Fat grams
sodium	Sodium millimeters
fiber	Fiber
carbo	Carbohydrates
sugar	Sugar
Potass	Potassium
vitamin	Vitamins
shelf	Shelf position in store
1 = bottom	
2 = middle	
3 = top	
weight	Weight (grams)
cups	Cups in serving
rating	Taste rating

CLT 100 random samples of size 50 from uniform distribution (*n* = 100). Data generated by SPSS.

Variable Name	**Variable Description**
u1	Results of random sampling, time 1
u2	Results of random sampling, time 2
.	
.	
.	
u100	Results of random sampling, time 100

CONFORM Husbands and wives conformity ratings (*n* = 20). Hypothetical data.

Variable Name	Variable Description
husband	Husband's score
wife	Wife's score

CRIME Crime rates per 100,000 for several types of crimes, by state (*n* = 50). Friendly, M. (1999). *Psych6140 Example SAS Programs.* Available: http://www.psych.yorku.ca/friendly/lab/files/psy6140/examples/factor/pca2.sas

Variable Name	Variable Description
murder	Murder rate
rape	Rape rate
robbery	Robbery rate
assault	Assault rate
burglary	Burglary rate
larceny	Larceny rate
auto	Automobile crime rate
state	State abbreviation

DEATH Data on number of months before, during, or after birth month that death occurred (*n* = 348). Philips, D. (1972). Deathday and birthday: An unexpected connection. In J.M. Tanner, *et al.* (Eds.), *Statistics: A guide to the unknown.* San Francisco: Holden Bay.

Variable Name	Variable Description
month	Month of death – month of birth

DELINQ Data on SES, population density, and delinquency for 75 community areas of Chicago (*n* = 75). Hypothetical data suggested by Galle, O.R., Gove, W.R., & McPherson, J.M. (1972). Population density and pathology: What are the relations for man? *Science, 176*, 23–30.

Variable Name	Variable Description
ses	Socioeconomic status (SES)
1 = low	
2 = high	
pop_dens	Population density
1 = low	
2 = high	
delinq	Delinquency
1 = low	
2 = high	

ENROLL Data on school districts, including the racial disproportion in classes for emotionally disturbed children (*n* = 26). U.S. Department of Education, Office for Civil Rights.

Variable Name	Variable Description
enroll	District enrollment
pct_aa	Percentage of students who are African-American
pct_lnch	Percentage of students who pay full-price for lunches
rac_disp	Racial disproportion in classes for emotionally disturbed[*]

*Positive index indicates that proportion of African-American students is greater than the
proportion of white students.

FIRE Data for 28 firefighter applicants (*n* = 28). Buffalo, New York, records.

Variable Name	Variable Description
candnum	Candidate ID number
sex	Sex
1 = male	
2 = female	
race	Race/ethnicity
1 = white	
2 = minority	
stair	Stair climb time (seconds)
body	Body drag time
obstacle	Obstacle course time
agility	Agility score
written	Written score
composite	Composite score

FOOTBALL Data on NFL football games for a recent year (*n* = 250). Publicly kept records.

Variable Name	Variable Description
date	Date of game
aw_tm	Name of away (visiting) team
aw_pt	Number of points away (visiting) team scored
ho_tm	Name of home team
ho_pt	Number of points scored by home team
predptsp	Predicted point spread
predou	Predicted total points
totpnts	Actual total points
actptsp	Actual point spread (winner points – loser points)

favored	Favored team
1 = home	
2 = away	
winner	Winning team
1 = home	
2 = away	
actou	Actual total points compared to predicted total points
0 = even	
1 = actual over the predicted	
2 = actual under the predicted	
winby	Points won by

HOTDOG Nutritional information for different brands of hot dogs (n = 54). Available through the Data and Story Library: http://lib.stat.cmu.edu/DASL/Datafiles/Hotdogs.html

Variable Name	Variable Description
type	Type of meat
1 = beef	
2 = other type of meat	
3 = poultry	
calories	Calories
sodium	Sodium millimeters

IQ IQ Scores for 23 children (n = 23) Anderson, T.W., & Finn, J.D. (1996). *The new statistical analysis of data.* New York: Springer-Verlag.

Variable Name	Variable Description
lang	Language IQ score
nonlang	Nonlanguage IQ score

IQ2 IQ Scores for 24 children (n = 24). Anderson, T.W., & Finn, J.D. (1996). *The new statistical analysis of data.* New York: Springer-Verlag.

Variable Name	Variable Description
lang	Language IQ score
nonlang	Nonlanguage IQ score

LIBRARY Size of book collection and number of staff for 22 college libraries (*n* = 22). McGrath, W.E. (1986). Levels of data in the study of library practice: Definition, analysis, inference and explanation. In G. G. Allen & F. C. A. Exon (Eds.), *Research and the practice of librarianship: An international symposium* (pp. 29–40). Perth, Australia: Western Australian Institute of Technology.

Variable Name	Variable Description
volumes	Number of volumes (100,000's)
staff	Number of staff

MOVIES Genre and gross for 100 top movies in 2001 (*n* = 100). Publicly held records.

Variable Name	Variable Description
movie	Movie name
reldate	Release date
opening	Opening week gross (millions)
total	Total gross (millions)
numtheat	Number of theatres in which the movie was released
weekstop	Number of weeks the movie was in the top 60
genre	Genre of movie

    1 = thriller/horror
    2 = family
    3 = drama
    4 = comedy
    5 = adventure/fantasy

NOISE Average highway speed and noise level for 30 sections of highway (*n* = 30). Hypothetical data suggested by Drew, D.R., & Dudek, C.L. (1965). *Investigation of an internal energy model for evaluating freeway level of service.* College Station: Texas A&M University, Texas Transportation Institute.

Variable Name	Variable Description
speed	Acceleration speed (mph)
noise	Noise level

POPULAR Data on elementary school students' goals (*n* = 478). Available through the Data and Story Library: http://lib.stat.cmu.edu/DASL/Datafiles/PopularKids.html

Variable Name	Variable Description
gender	Gender
grade	Grade

    1 = female
    2 = male

age	Age
race	Race/ethnicity
1 = white	
2 = other	
urban	School urbanicity
1 = rural	
2 = suburban	
3 = urban	
school	School name
goals	Student goals
1 = make good grades	
2 = be popular	
3 = be good at sports	
grades	Importance of grades for popularity (1 = most; 4 = least)
popular	Importance of sports for popularity (1 = most; 4 = least)
looks	Importance of looks for popularity (1 = most; 4 = least)
money	Importance of money for popularity (1 = most; 4 = least)

READING Reading scores of 30 students before and after second grade (n = 30). Records of a second-grade class.

Variable Name	Variable Description
Before	Reading test score before second grade
After	Reading test score after second grade

SLEEP Data on mammals' physical, environmental, and sleep characteristics (n = 62). Available: http://lib.stat.cmu.edu/datasets/sleep

Variable Name	Variable Description
species	Species' name
bodywt	Body weight (kg)
brainwt	Brain weight (g)
nodream	Non-dreaming sleep (hrs/day)
dream	Dreaming sleep (hrs/day)
totsleep	Total sleep (hrs/day)
lifespan	Life span (years)
gestate	Gestation time (days)
prey	Predation index (1 = min to 5 = max)
sleepexp	Sleep exposure index (1 = least exposed to 5 = most exposed)

danger

Danger index – combination of
predation and sleep exposure indices (1 =
least to 5 = most)

Missing values = –999

SOCMOB Data on family structure and occupation of members (*n* = 1156).Data from: Biblarz, T.J., & Raftery, A.E. Raftery. (1993). The Effects of Family Disruption on Social Mobility. *American Sociological Review*. Data available from StatLib: http://lib.stat.cmu.edu/datasets/socmob

Variable Name	**Variable Description**
Idnum	Family identification number
f_occup	Father's occupation
1 = laborer	
2 = craftsperson	
3 = salesperson	
4 = manager	
5 = professional	
family	Family structure
1 = intact	
2 = non-intact	
race	Race
1 = white	
2 = other	
s_occup	Son's occupation
1 = laborer	
2 = craftsperson	
3 = salesperson	
4 = manager	
5 = professional	

SPIT Data on success of interventions to curb chewing spitting tobacco (*n* = 54). Greene,J.C., Walsh, M.M., & Mosouredis, C. (1994). Report of a pilot study: A program to help major league baseball players quit using spit tobacco. *Journal of the American Dental Association, 125*, 559-567.

Variable	**Variable Name**
interven	Type of intervention
1 = Minimum	
2 = Extended	
outcome	Outcome of intervention
1 = Subject quit entirely	
2 = Subject tried unsuccessfully to quit	
3 = Subject failed to try to quit	

STEPPING Information on heart rate after stepping exercise, based on differences in stepping frequency and step height ($n = 30$). Available through the Data and Story Library: http://lib.stat.cmu.edu/DASL/Datafiles/Stepping.html

Variable Name	Variable Description
Order	Order in study
Block	Subject and experimenter block ID number
Height	Height range of step
0 = low (5.75inches)	
1 = high (11.5 inches)	
Frequency	Frequency of stepping
0 = slow	
1 = medium	
2 = fast/high	
Resthr	Resting heart rate, beats per minute
Hr	Heart rate after exercise, beats per minute

TITANIC Sex, age, and survival outcome for passengers on Titanic. ($n = 2201$). Available through Journal of Statistics Education Data Archive:
http://www.amstat.org/publications/jse/datasets/titanic.dat

Variable Name	Variable Description
class	Classification - passenger class and crew
0 = crew	
1 = first class	
2 = second class	
3 = third class	
age	Age level
0 = child	
1 = adult	
sex	Sex
0 = female	
1 = male	
survived	Survival status
0 = no	
1 = yes	

WAR Expectations of possibility of war ($n = 597$). Lazarsfeld, P.F., Berelson, B., & Gaudet, H. (1968). The People's Choice (3rd edition). New York: Columbia University Press.

Variable Name	Variable Description
June	Response in June 1948
0 = does not expect war	
1 = expects war	

October   Response in October 1948
    0 = does not expect war
    1 = expects war

WEATHER Average precipitation and temperature on July 2nd for U.S. cities (n = 78). Data obtained from the Internet.

Variable Name	Variable Description
City	Name of city
Temp	Temperature (degrees Fahrenheit)
Precip	Inches of rainfall

WORDS Number of words 18 children memorized based on three different experimental conditions (n = 18). Hypothetical data.

Variable   **Variable Name**

info_set   Information set
    1 = no information
    2 = "3 categories"
    3 = "6 categories"
words   Number of words memorized

Appendix B

Answers to Selected Chapter Exercises

Chapter 1

1.1

a.   13 variables.

b.   String, 10 characters.

1.2

a.   No, the format is an SPSS data file, not an ASCII data file.

b.   54 cases.

c.   2 variables.

d.   No, there are no missing data.

Chapter 2

a.   The frequency distribution of "temp" is given in the table (following page). 10 adults have a body temperature of 98.6°.

b.   62.3% of the adults in the sample have a body temperature less than 98.6° (that is, 98.5° or less).

body temperature (degrees Fahrenheit)

		Frequency	Percent	Valid Percent	Cumulative Percent
Valid	96.3	1	.8	.8	.8
	96.4	1	.8	.8	1.5
	96.7	2	1.5	1.5	3.1
	96.8	1	.8	.8	3.8
	96.9	1	.8	.8	4.6
	97.0	1	.8	.8	5.4
	97.1	3	2.3	2.3	7.7
	97.2	3	2.3	2.3	10.0
	97.3	1	.8	.8	10.8
	97.4	5	3.8	3.8	14.6
	97.5	2	1.5	1.5	16.2
	97.6	4	3.1	3.1	19.2
	97.7	3	2.3	2.3	21.5
	97.8	7	5.4	5.4	26.9
	97.9	5	3.8	3.8	30.8
	98.0	11	8.5	8.5	39.2
	98.1	3	2.3	2.3	41.5
	98.2	10	7.7	7.7	49.2
	98.3	5	3.8	3.8	53.1
	98.4	9	6.9	6.9	60.0
	98.5	3	2.3	2.3	62.3
	98.6	10	7.7	7.7	70.0
	98.7	8	6.2	6.2	76.2
	98.8	10	7.7	7.7	83.8
	98.9	2	1.5	1.5	85.4
	99.0	5	3.8	3.8	89.2
	99.1	3	2.3	2.3	91.5
	99.2	3	2.3	2.3	93.8
	99.3	2	1.5	1.5	95.4
	99.4	2	1.5	1.5	96.9
	99.5	1	.8	.8	97.7
	99.9	1	.8	.8	98.5
	100.0	1	.8	.8	99.2
	100.8	1	.8	.8	100.0
	Total	130	100.0	100.0	

c.   The lowest temperature is 96.3°; the highest is 100.8°.

d.   The histogram of "temp" with 12 intervals is listed below:

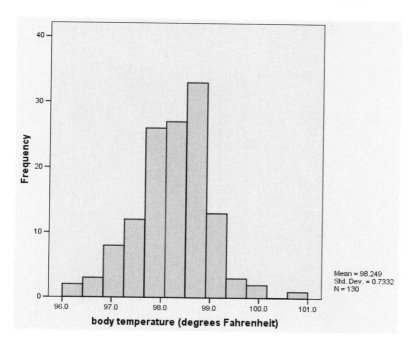

2.2

a.   There are 14 male firefighter applicants, which is 50% of the sample.

b.   39.3% of the applicants are minorities.

c.   64.3% of the women scored below 18 seconds in the stair climb task.

d.   100% of the men scored below 18 seconds on the stair climb task.

e.   The stem-and-leaf plot of the written test score:

```
        WRITTEN Stem-and-Leaf Plot

        Frequency     Stem &   Leaf

            7.00        7 .  0011234
            9.00        7 .  555678889
            3.00        8 .  003
            4.00        8 .  6668
            3.00        9 .  023
            2.00        9 .  58

        Stem width:   10.00
        Each leaf:       1 case(s)
```

f.   4 applicants had scored between 85 and 89 on the written test.

2.3

a. Bar chart of the "class" variable is listed below. The crew level had the most passengers.

b. There were more first-class passengers than second-class passengers.

c. 711 passengers survived.

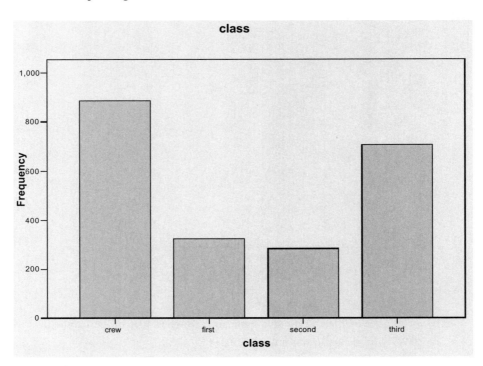

Chapter 3

3.1

a. mean = 373.4091, median = 346. The differences may be due to the shape of the distribution. Because the mean is larger than the median, the distribution is most likely positively skewed.

b. The 10th percentile is 232.2; the 90th is 634.7.

c. The histogram is displayed below. The histogram shows the skewness of the distribution. The libraries with over a 600 person staff (specifically, those with 677 and 711 staff members) appear to be outliers.

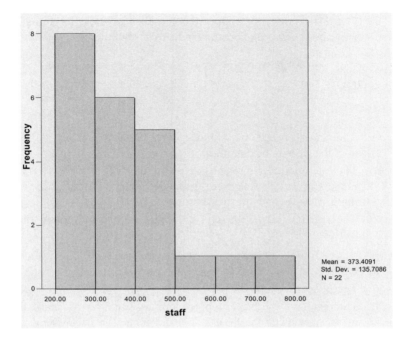

3.2

a. The mean is 114.53 seconds.

b. The mean is 116.53 seconds.

c. Part (b) is 2 seconds greater than part (a). The general rule is: When you add a constant to each observation in a data set, the mean of the transformed data is equal to the original mean plus the constant.

d. The mean of the new data is 57.26 seconds, or the original mean divided by 2.

3.3

a. The mean, median, and mode are contained in the Frequencies table below. Note that only one mode for language IQ is listed. There are five other modes — 94, 95, 99, 102, and 105.

Statistics

		language IQ	nonlanguage IQ
N	Valid	23	23
	Missing	0	0
Mean		97.57	50.30
Median		96.00	47.00
Mode		86[a]	43

a. Multiple modes exist. The smallest value is shown

b. The distributions for both of the variables are close to normal, so the mean is the best measure of central tendency for both.

Chapter 4

4.1

a. The range is 5,200,000 volumes, or 52 (when expressed in 100,000 volumes as is the case in the data file).

b. The interquartile range (expressed in 100,000) is 19.975. The interquartile range is a better measure of dispersion than is the range in instances when there are outliers in the distribution.

c. The standard deviation is 14.52879 (100,000 volumes); the variance is 211.086.

4.2

a. The z-score for *Cats and Dogs* is .16

b. The results of the Explore procedure are listed below, and indicate that the movie that made the least amount of money in its opening week had a z-score of −1.05; the one that made the most had a z-score of 4.00.

Descriptives

			Statistic	Std. Error
Zscore: Opening week gross	Mean		.0000000	.10000000
	95% Confidence Interval for Mean	Lower Bound	-.1984217	
		Upper Bound	.1984217	
	5% Trimmed Mean		-.1130548	
	Median		-.3057197	
	Variance		1.000	
	Std. Deviation		1.000000	
	Minimum		-1.05058	
	Maximum		3.99773	
	Range		5.04831	
	Interquartile Range		.6890511	
	Skewness		1.933	.241
	Kurtosis		3.740	.478

4.3

a. The standard deviation of agility score is 5.72 seconds.

b. The standard deviation does not change when a constant is subtracted from each time.

c. The standard deviation is halved when each person's time is halved.

4.4

a.   The results of the Explore procedure are displayed below:

Descriptives

type of hot dog				Statistic	Std. Error
CALORIES	beef	Mean		156.8500	5.06291
		95% Confidence Interval for Mean	Lower Bound	146.2532	
			Upper Bound	167.4468	
		5% Trimmed Mean		157.5556	
		Median		152.5000	
		Variance		512.661	
		Std. Deviation		22.64201	
		Minimum		111.00	
		Maximum		190.00	
		Range		79.00	
		Interquartile Range		40.2500	
		Skewness		-.031	.512
		Kurtosis		-.813	.992
	meat	Mean		158.7059	6.12058
		95% Confidence Interval for Mean	Lower Bound	145.7308	
			Upper Bound	171.6809	
		5% Trimmed Mean		159.5621	
		Median		153.0000	
		Variance		636.846	
		Std. Deviation		25.23580	
		Minimum		107.00	
		Maximum		195.00	
		Range		88.00	
		Interquartile Range		42.0000	
		Skewness		-.209	.550
		Kurtosis		-.823	1.063
	poultry	Mean		118.7647	5.46952
		95% Confidence Interval for Mean	Lower Bound	107.1698	
			Upper Bound	130.3596	
		5% Trimmed Mean		118.7386	
		Median		113.0000	
		Variance		508.566	
		Std. Deviation		22.55141	
		Minimum		86.00	
		Maximum		152.00	
		Range		66.00	
		Interquartile Range		42.0000	
		Skewness		.025	.550
		Kurtosis		-1.605	1.063

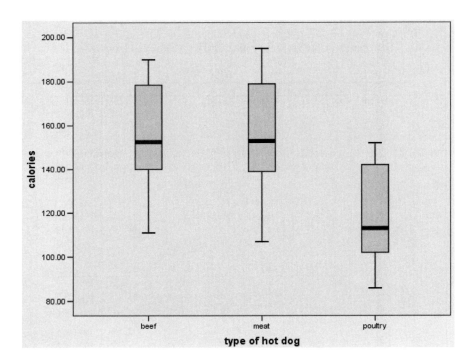

b. The median is 152.5 calories for beef, 153.0 calories for meat, and 113.0 calories for poultry.

c. The minimum and maximum number of calories for beef hot dogs is 111 calories to 190 calories, respectively.

d. There are no outliers for poultry hot dogs; the box-and-whisker plot shows that there are no stray points above or below the whiskers.

e. The meat hot dogs have the most variability (the largest standard deviation).

Chapter 5

5.1

a. The Pearson correlation is -.410

5.2

a. The Pearson correlation is .926

b. The Spearman correlation is .833

c. Both coefficients indicate a fairly strong, positive association between exposure and mortality.

5.3

The correlation matrix is displayed below:

Correlations

		ENROLL	PCT_AA	PCT_LNCH	RAC_DISP
ENROLL	Pearson Correlation	1	-.491*	.089	.204
	Sig. (2-tailed)	.	.011	.666	.319
	N	26	26	26	26
PCT_AA	Pearson Correlation	-.491*	1	-.644**	-.431*
	Sig. (2-tailed)	.011	.	.000	.028
	N	26	26	26	26
PCT_LNCH	Pearson Correlation	.089	-.644**	1	.467*
	Sig. (2-tailed)	.666	.000	.	.016
	N	26	26	26	26
RAC_DISP	Pearson Correlation	.204	-.431*	.467*	1
	Sig. (2-tailed)	.319	.028	.016	.
	N	26	26	26	26

*. Correlation is significant at the 0.05 level (2-tailed).

**. Correlation is significant at the 0.01 level (2-tailed).

a. The correlation between percentage of students who pay full price for lunches and percent of student who are African-American is largest in magnitude (-.644).

b. The negative correlation between "enroll" and "pct_aa" indicates that districts with low enrollment tend to have a high percentage of students who are African-American.

c. Racial disproportion is most highly strongly associated with percentage of students who pay full price for lunch. It is positive and moderately strong (.467).

Chapter 6

6.1

a. 62.5% of first class passengers survived; 25.2% of third class passengers survived.

b. There were 885 crew on board; 24.0% of them survived.

c. It seems that the first class passengers were more likely to survive than either the crew or third class passengers.

6.2

a. 21.4% of the applicants were minority females.

b. White – 9 (or 64.3%) compared to 5 (or 35.7%).

c. Phi is .073, a very weak association.

6.3

a. For both boys and girls, making good grades is the most popular goal (selected by 51.8% of females and 51.5% of males). Being good in sports is more important to boys (26.4%) than to girls (12.0%). Similarly, being popular is somewhat more important to girls than to boys (36.3% compared to 22.0%).

b. The pattern described in part (a) is true in both suburban and urban schools, but not in rural schools. In rural locations, among boys, being good in sports is slightly more popular than is making good grades; and among girls, being popular is almost as important as is making good grades.

Chapter 8

8.1 z = -2.

8.2 50%.

8.3 z = .25.

8.4 84%.

8.5

a. 84%.

b. 95.4%.

Chapter 10

10.1

a. t = .482, $P < .633$. Accept H_0. Conclude that average speed of vehicles is not different from 35 mph.

b. No.

10.2

a. Sample mean = 11.72 pints; 90% confidence interval: (10.80, 12.65)

b. Yes, reject the null hypothesis because 10 points is not within the confidence interval.

c. Reject the null hypothesis and conclude that games are won by, on average, more than 10 points.

d. The two-tailed P value is $P < .002$ so the one-tailed P is $P < .001$.

10.3

a. $P < .0005$.

b. Reject at $\alpha = .05$ and at $\alpha = .01$.

10.4

a. 51.7%

b. Do not reject the null hypothesis ($P < .493$).

10.5

a. Yes, wives are more conformist, on average, than their husbands.

b. The minimum α for rejecting the null hypothesis is .012 (because this is a one-tailed test).

Chapter 11

11.1

b. Perform a one-tailed test because there is reason to believe that the racial disproportion will be greater in areas that are "low" in terms of percentage of students who pay full price for lunch.

c. $H_0 : \mu_{low} \leq \mu_{high}$ $H_1 : \mu_{low} > \mu_{high}$

d. $t = -1.970$; $P < .060$ for a two-tailed test, so for the one-tailed test it is $P < .030$. Reject H_0 at .05, but not at .01.

11.2

a. $H_0 : \mu_{home} \leq \mu_{away}$ $H_1 : \mu_{home} > \mu_{away}$

b. The P value is .033 (because this is a one tailed test, we divide .066 by 2). Because .033 is not less than .01, we do not reject the null hypothesis, and we conclude that when home teams win, it is not by more points than when away teams win.

11.3

a. $H_0 : \mu_{students} = \mu_{faculty}$ $H_1 : \mu_{students}$ $\mu_{faculty}$

b.   The P value is less than .0005, which is less than our significance level, so we concluded that, on average, students drive cars that are older than do faculty (in the sample, the average is 6.25 years for students compared to 3.82 years for faculty).

Chapter 12

12.1

a. $\chi^2 = 349.915$, $P < .0005$, conclude that class and survival are not independent.

b.   Percentages by class indicate that first class passengers were more likely to survive.

12.2

a.   Sex and views on the importance of money are independent. $\chi^2 = 2.761$, $P < .430$.

b.   Sex and views of the importance of looks are not independent. $\chi^2 = 77.059$, $P < .0005$. The percentages by sex indicate that looks are more important to girls (in the sample, 56.2% of girls rated it most important, compared to 19.4% of boys).

12.3

a.   If we assume that both the variables are ordinal, and that intervention type will predict outcome, then the appropriate measure is Somer's d.

b.   The value of the correlation coefficient is $-.567$, with $P < .0005$. The P value is less than our significance point, so we conclude that the variables are related. The strength of the association is moderate. The negative sign indicates that the extended intervention (which is coded 1 on a 0,1 scale) tends to be related to successful quitting (which is coded 1, one a 1–3 scale).

Chapter 13

13.1

a.   Scatter plot with regression line:

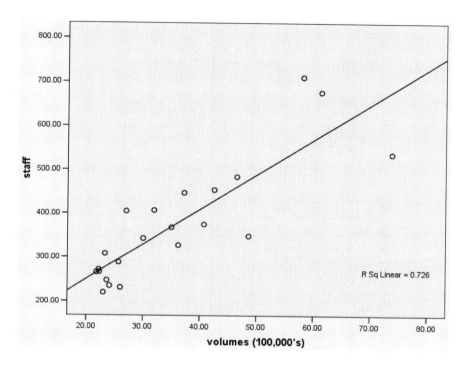

b. The relationship is positive.

c. The relationship appears strong.

13.2

b. The slope, β, is –2.401, standard error = .237, 95% confidence interval for β (–2.874, –1.928).

c. There is a significant relationship between sugar and rating. $t = -10.117$, $P < .0005$. The relationship indicates that cereals with comparatively larger amounts of sugar are rated better tasting (because rating is scored as 1= best tasting).

d. $r = .760$.

e. .577. 57.7% of the variation in rating of breakfast cereals is attributed to differences in sugar content.

13.3

c. Yes, there is a significant relationship between gender and agility; men are more agile than women (lower agility scores are superior). This can be discerned in several ways. The simple correlation is –.766, $P < .0005$. The

F-test for the Model is F = 36.943, $P < .0005$. The t-test for β is $t = -.6078$, $P < .0005$.

d. Mean for men = $-.7464$; mean for women = 6.0736. The difference is -6.82, which is equivalent to the regression coefficient.

e. The raw coefficient. When the independent variable is a dummy variable, a "one unit increase" is equivalent to moving form one group to the other. Thus, the raw regression coefficient is always equal to the mean difference between groups.

13.4

a. Yes, the overall F = 10.259, and $P < .0005$.

b. Exposure of the den is related to hours of dream sleep per day controlling for prey index ($t = -2.646$, $P < .011$). The relationship between exposure of the den and amount of dream sleep is such that each unit increase in exposure of an animal's den (on a 5-point scale) is associated with .397 fewer hours of sleep per day. The likelihood of being preyed upon is not related to amount of dream sleep after controlling for den exposure ($t = -1.016$, $P < .315$).

c. $R^2 = 30.4\%$.

d. Overall, there is a relationship among the variables (F = 7.110, $P < .001$). The proportion of variance accounted for increases to 31.7%. After controlling for body weight, predation index is not related to dream sleep but den exposure is. The magnitude of the relationship of den exposure increased from the first model, that is, a one-unit increase in exposure is associated with .468 fewer hours of sleep per day.

Chapter 14

14.1

a. F = 16.704, $P < .0005$. Conclude that different types of hot dogs have different average calories.

b. Poultry hot dogs have fewer calories than either beef ($P < .0005$) or meat ($P < .0005$) hot dogs. There is no significant difference between meat and beef hot dogs ($P < 1.000$). Means in the sample are beef = 156.85 calories, mean = 158.71 calories, poultry = 118.76 calories.

c. Effect size for poultry with beef is 1.62 standard deviations; for poultry with meat it is 1.70 standard deviations.

d. F = 1.778. Different types of hot dogs do not differ, on average, in amount of sodium.

14.2

a.   F = 1.867, $P < .123$. Conclude that genre does not affect the number of weeks movies stay in the Top 60.

b.   Because there is no statistically significant difference, it is not appropriate to conduct post hoc tests.

c.   No post hoc tests were conducted.

Chapter 15

15.1

a.   76.81%

b.   Two components had eigenvalues greater than 1.

c.   Because we used the default extraction criteria, the two components with eigenvalues greater than 1 were extracted.

d.   The variables: larceny, auto, burglary and robbery load heavily on the first (rotated) component; murder, assault, and rape load heavily on the second (rotated) component. Thus, the first component may be said to represent "property crime" and the second "personal injury crime."

15.2

a.   .9486

b.   It decreases to .9417

15.3

The coefficient alpha is .5367, which is somewhat low. If the obstacle course time is deleted, however, the coefficient alpha increases to .9097.

Index

α, alpha
    coefficient, 168
    error, 101
    intercept, 124
β, beta, 124
χ^2, chi, 117
γ, gamma, 119
λ, lambda, 119
φ, phi, 72, 119
μ, mu, 101
σ, sigma, 98

ASCII, 5, 9
Alternative hypothesis
    (H_1), 99
Analysis of variance
    (ANOVA), 144, 155,
    157, 163, 164, 171, 173

b, least-squares estimate of
    regression coefficient,
    128
Bar chart, 33
Bernoulli distribution, 82
Binomial procedure, 103
Bonferroni, 147
Box-and-whisker plot,
    56–57

Central limit theorem, 95,
    102

Change statistics, 137
Chi-square, 108, 114, 117
Confidence interval, 110,
    115
Continuity correction, 104
Correlation coefficient,
    69–70, 140
Correlation matrix, 71–73
Cross-tabulation, 69
Cumulative percent, 29

Data
    categorical, 27
    continuous, 27
    discrete, 27
    numerical, 27
Data reduction, 160
Deletion
    listwise, 19, 65
    pairwise, 19, 65
Descriptive statistics
    procedure, 28
Descriptives, 45

Effect size, 171
Excel (Microsoft), 9
Explore procedure, 33

F statistic, 127, 146
Frequency distribution, 27

225